D1738817

DOES DEBT MANAGEMENT MATTER?

The Trade Union Institute for Economic Research, FIEF, is a foundation established in 1985 by Landsorganisationen, the Swedish trade union confederation. FIEF's objective, as defined in its constitution, is to 'deepen the academic economic debate through the promotion of enduring research'.

FIEF Studies in Labour Markets and Economic Policy will be published once a year. The series will provide a forum for outstanding scholars to publish applied, policy-oriented research with generous space available. The length of the papers should be between 40 and 60 pages which allow background surveys of theory, and a review of empirical research. The papers should also contain original contributions either through extensions and/ or of empirical analysis.

Normally, two conferences are organized around the papers to be published in the *FIEF Studies*. After the first conference, papers are revised and a final conference is held with FIEF's panel of advisors and specially invited researchers in the field covered by the paper.

FIEF Studies editorial board

Managing Editor:
Villy Bergström   Director of the Trade Union Institute for Economic Research, FIEF

Editorial Board:
Lars Calmfors   Professor, Institute for International Economic Studies, IIES, University of Stockholm

Michael Hoel   Professor, Oslo University
Bertil Holmlund   Professor, Uppsala University
Karl-Gustaf Löfgren   Professor, University of Umeå
Andrew Oswald   Dr, The London School of Economics
Hans T. Söderström   Executive Director, Center for Business and Policy Studies, SNS

FIEF panel for this volume

Jonas Agell   Docent, Institute for International Economic Studies (IIES), University of Stockholm

Dan Andersson	Economist, Swedish Trade Union Confederation, LO
Peter Englund	Professor, Uppsala University
Jeffrey A. Frankel	Professor, University of California, Berkeley
Benjamin M. Friedman	Professor, Harvard University
Per-Olov Johansson	Professor, Stockholm School of Economics
Björn Järnhäll	Economist, Bank of Sweden
Urban Karlström	Dr, The Swedish National Debt Office
Karl-Gustaf Löfgren	Professor, Umeå University
Christian Nilsson	Economist, Bank of Sweden
Anders Paalzow	MBA, Stockholm School of Economics
Mats Persson	Professor, Institute for International Economic Studies, IIES, University of Stockholm
Torsten Persson	Professor, Institute for International Economic Studies, IIES, University of Stockholm
Eva Srejber	Economist, Bank of Sweden
Lars Werin	Professor, University of Stockholm
Staffan Viotti	Professor, Stockholm School of Economics
Anders Vredin	Dr, Trade Union Institute for Economic Research

Does Debt Management Matter?

Jonas Agell
Mats Persson
and
Benjamin M. Friedman

CLARENDON PRESS · OXFORD
1992

Oxford University Press, Walton Street, Oxford OX2 6DP
Oxford New York Toronto
Delhi Bombay Calcutta Madras Karachi
Petaling Jaya Singapore Hong Kong Tokyo
Nairobi Dar es Salaam Cape Town
Melbourne Auckland
and associated companies in
Berlin Ibadan

Oxford is a trade mark of Oxford University Press

Published in the United States
by Oxford University Press, New York

© *FIEF 1992*

All rights reserved. No part of this publication may be reproduced,
stored in a retrieval system, or transmitted, in any form or by any means
electronic, mechanical, photocopying, recording, or otherwise, without
the prior permission of Oxford University Press

British Library Cataloguing in Publication Data
Data available

Library of Congress Cataloging-in-Publication Data
Agell, Jonas.
Does debt management matter?/Jonas Agell, Mats Persson, and
Benjamin M. Friedman.
p.   cm. — (FIEF studies in labour markets and economic
policy)
Includes bibliographical references and index.
1. Corporate debt.   I. Persson, Mats, 1949–   II. Friedman,
Benjamin M.   III. Title.   IV. Series.
HG4028.D3A46   1992   658.15'26—dc20   91–32815
ISBN 0–19–828361–X

Typeset by Best-set Typesetter Ltd., Hong Kong
Printed in Great Britain by
Bookcraft Ltd, Midsomer Norton, Avon

Contents

List of Figures	ix
List of Tables	x
Introduction VILLY BERGSTROM	1

Part I: Does Debt Management Matter? — 5
JONAS AGELL AND MATS PERSSON

1. Introduction	7
2. Some General Concepts	12
3. The Portfolio Balance Approach to Debt Management	32
4. Implementing the Basic Model by Using Historical Data	38
5. An Alternative Approach to the Covariance Matrix	59
6. How Returns Adjust: The Effects of Endogenous Prices	66
7. Summary and Conclusions	78
8. Comment	81
JEFFREY A. FRANKEL	
9. Comment	92
BENJAMIN M. FRIEDMAN	
References for Part I	103

Part II: Debt Management Policy, Interest Rates, and Economic Activity — 109
BENJAMIN M. FRIEDMAN

10. Introduction	111
11. Debt Management, Interest Rates, and Asset Prices	115
12. A Model of Interest Rates and Economic Activity	123
13. Empirical Assessment of the Effects of Debt Management Policies	129
14. Summary and Conclusions	140

viii   Contents

15. Comment   141
STAFFAN VIOTTI

References for Part II   146

Index   149

List of Figures

1.1 Gross public debt as a percentage of GNP for five OECD
countries, 1972–1986   8

4.1 A deterministic cycle with random disturbances   42

4.2 Conditional quarterly covariances between the real yields
on (1) corporate stock, (2) long-term bonds, and (3)
short-term bonds, 1970(1)–1988(2)   46

4.3 Conditional coefficients of correlation between the
quarterly yields on (1) corporate stock, (2) long-term
bonds, and (3) short-term bonds, 1970(1)–1988(2)   47

4.4 The effects on asset yields of increasing the supply of
long-term bonds: quarterly data, 1970(1)–1988(2)   48

4.5 The realization of two hypothetical stochastic processes   53

4.6 The covariances of the two time series in Fig. 4.5 for
different data intervals   54

4.7 Conditional monthly covariances between the real yields
on (1) corporate stock, (2) long-term bonds, and (3) short-
term bonds, 1970(1)–1988(2)   56

4.8 Conditional coefficients of correlation between the real
yields on (1) corporate stock, (2) long-term bonds, and (3)
short-term bonds: monthly data, 1970(1)–1988(2)   57

4.9 (a) The derivative $\partial r_1^c/\partial \alpha_2^s$ for monthly and quarterly
data, 1970–1988   58

    (b) The derivative $\partial r_2^c/\partial \alpha_2^s$ for monthly and quarterly
data, 1970–1988   58

5.1 A comparison of the quarterly variances of nominal
asset yields using options data versus autoregression
procedures, 1985(4)–1988(2)   62

5.2 (a) The derivative $\partial r_1^c/\partial \alpha_2^s$ for options and VAR
approaches: quarterly data, 1985(4)–1988(2)   64

    (b) The derivative $\partial r_2^c/\partial \alpha_2^s$ for options and VAR
approaches: quarterly data, 1985(4)–1988(2)   64

6.1 (a) The derivative $\partial r_1^c/\partial \alpha_2^s$ for models with and without
valuation changes: quarterly data, 1970(1)–1988(2)   77

    (b) The derivative $\partial r_2^c/\partial \alpha_2^s$ for models with and without
valuation changes: quarterly data, 1970(1)–1988(2)   77

8.1 Covered interest differential (3-month local Eurodollar)   90

13.1 Simulated effects of two debt management actions   136

List of Tables

1.1 Average of remaining years to maturity of public debt outstanding, 1980–1987   9

4.1 Covariations between the quarterly real yields of (1) corporate equity, (2) long-term government bonds, and (3) short-term government bonds: US data, 1960(1)–1988(2)   40

4.2 Covariations between the quarterly nominal yields of (1) corporate equity, (2) long-term government bonds, and (3) the consumer price index: US data, 1960(1)–1988(2)   40

4.3 Conditional covariances between the quarterly real yields of (1) corporate equity, (2) long-term government bonds, and (3) short-term government bonds: US data, 1960(1)–1988(2)   44

10.1 Maturity structure of marketable interest-bearing US Treasury debt held by private investors, 1945–1990   112

13.1 Simulated effects of two debt management actions   130

Introduction

In the mid-1970s, after the first oil crisis, many countries began to run larger deficits on government budgets than earlier during the postwar period. For instance, the government debt in Japan and Sweden increased from 20–30 per cent of GNP in the early 1970s to around 70 per cent in the 1980s. As another example, in the early 1980s the US government deficit increased, as a result of the Reagan tax policy, by so much that the public debt went up some 10 percentage points of GNP to 50 per cent towards the end of the decade.

This growth of government debt has occurred concomitantly with a development, liberalization, and sophistication of capital markets. In fact, these latter events have probably been a prerequisite for the growing government indebtedness. In Sweden the debt management is run by a 200-year-old institution, the 'National Debt Office' (Riksgäldskontoret). This old agency has in the last decade emerged from oblivion. Its former staff of obscure 'petty functionaries' has changed and nowadays consists of policy stars, members of a jet set who pursue favourable borrowing conditions on the international capital markets.

The growth of public debt has stimulated the interest of academic economists. Does debt management matter? In recent years there has been a discussion of the debt burden of underdeveloped countries and of the neutrality of total government debt in more advanced economies. However, the possible effects of the management of a given debt on real capital formation via portfolio crowding-out or crowding-in has been relatively neglected. This is why Volume 3 of *FIEF Studies in Labour Markets and Economic Policy* is fully devoted to two studies of debt management.

The paper by Professors Jonas Agell and Mats Persson deals with a purely financial model, analysing the effects of debt management policies on relative asset yields. Professor Benjamin

2   *Villy Bergström*

Friedman takes this one step further. He raises the question of how debt management influences the 'real economy': real output, real capital formation, and consumption.

Agell and Persson carefully discuss how to define 'debt management' as opposed to monetary and fiscal policy. They derive a portfolio balance model of asset shares. This model is then used to study which changes in the composition of government debt can change the structure of asset yields. For instance, will a consolidation of the government debt change the long- and short-term rate of interest as well as the cost of capital to the business sector?

The important question regarding the possibilities of influencing relative asset yields is, to what extent are assets in the portfolios of investors close substitutes? This question hinges on the variance of asset yields and on the covariance between asset yields, conditional on the information that investors have at the time of portfolio investment. An important contribution in the Agell–Persson paper is their attempt at inferring, in different ways, the conditional variance–covariance structure perceived by investors in the capital market.

Inference about the variance–covariance structure perceived by investors will depend on the time perspective that the investors are believed to have. Agell and Persson therefore focus on the problem of aggregation over time by comparing quarterly and monthly time-series data. They then complicate their analysis by allowing the conditional variance–covariance structure in their model to change over time, as well as by endogenizing asset prices, by letting changes in the supply of assets affect expected returns.

There are two Comments on the Agell–Persson paper, one by Professor Jeffrey Frankel, University of California at Berkeley, and one by Professor Benjamin Friedman, Harvard University, who has also written the second paper for this FIEF volume.

The ultimate reason for interest in debt management is, after all, its possible effect on the real economy, i.e. on production, growth of production, and employment. Agell and Persson are reluctant to make judgements about real effects based on their analysis of financial markets. They consider the financial effects to be too small and too uncertain to allow inference about real variables.

Introduction

Benjamin Friedman's contribution to this volume marries the financial markets to the non-financial economy. Professor Friedman uses basically the same theoretical structure as Agell and Persson for the determination of interest rates in four sub-markets of the US government securities market. This model of interest determination is used in the MIT–Penn–SSRC (MPS) econometric model of the US economy. The term-structure equation of that model is replaced by the structural model of interest rate determination in Professor Friedman's paper.

The econometric macro model is used for two simulation experiments. The first is a sustained shift of new issues for Treasury financing from long-term securities to short-term securities. The second is a one-year programme to shorten the maturity structure of outstanding debt by purchasing long-term securities and issuing short-term securities. The magnitude of the changes in debt volumes are chosen to be reasonable in relation to the existing stock and flow of government debt. Friedman's analysis shows that a shortening of the maturity structure of the government debt by both examples might increase output, and that it would do so by spurring the rate of real capital formation. Professor Friedman's conclusion is that active debt management has effects on both interest rates and economic activity in the non-financial part of the economy. The magnitudes of the changes involved are large enough to make debt management worthy of further economic analysis.

The paper by Benjamin Friedman is commented on by Professor Staffan Viotti of the Stockholm School of Economics.

Villy Bergström
Director, FIEF

PART I

Does Debt Management Matter?

1

Introduction

The persistent deficits in government budgets over the last two decades have caused a substantial accumulation of public debt throughout the Western world. At the beginning of the 1970s the average ratio of gross public debt to GNP in the OECD was 38 per cent, whereas by 1986 it had increased to 56 per cent.[1] Behind these aggregate figures there are substantial differences both across time and across individual countries. For example, Japan experienced a dramatic increase from a very low level (17 per cent in 1972) to a very high level (67 per cent in 1986); the UK had a fall from a very high level (75 per cent) to an intermediate (55 per cent) level; while the USA experienced a decline from 44 per cent in 1972 to 37 per cent in 1981, followed by a fairly rapid increase to 48 per cent in 1986.[2] In Fig. 1.1 we show the historical development of the public debt–GNP ratio for a sample of OECD countries. Although these figures are not fully comparable, neither across countries nor over time for a single country, they indicate roughly the actual development.

These developments have led to a growing interest in the economics of deficit financing and the controllability of an increasingly complex and sophisticated financial system. The purpose of this paper is to discuss the theory and evidence of one particular economic policy towards financial markets, namely government debt management. In general terms, the question is to what extent changing the composition of public debt, rather than its size, can facilitate macroeconomic policy-making. The main channel for debt management policy is the portfolio choice

---

We are grateful for helpful comments and suggestions by Jeffrey A. Frankel, Benjamin M. Friedman, Staffan Viotti, and Lars Werin.

[1] All figures on debt–GNP ratios in this section are taken from Chouraqui *et al.* (1986: 108).
[2] Of all OECD countries, only the UK, Australia, and Norway experienced a declining debt–GNP ratio during the period.

8 *Jonas Agell and Mats Persson*

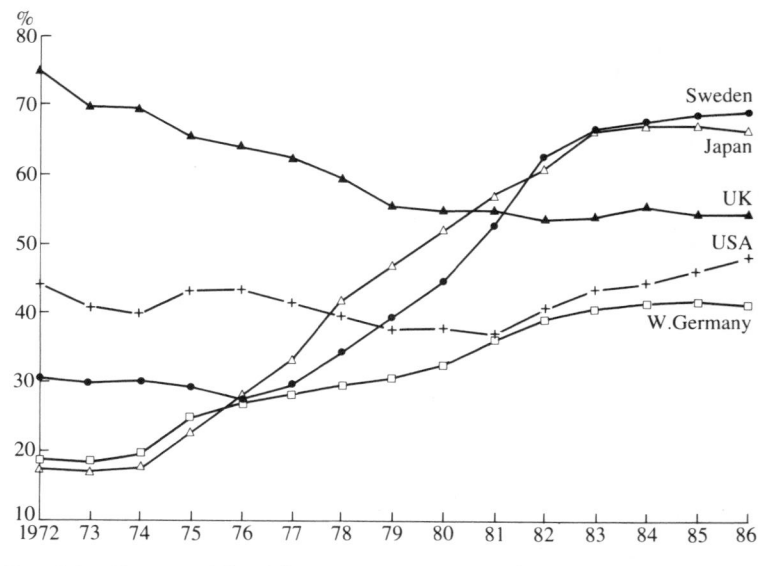

FIG. 1.1 Gross public debt as a percentage of GNP for five OECD countries, 1972–1986
(*Source*: Chouraqui *et al*. 1986)

of investors and the pricing of assets in financial markets. To the extent that different debt instruments are considered as less than perfect substitutes, changing the pattern of government borrowing will alter the structure of relative asset yields. This will in turn affect the 'real' side of the economy, since the consumption and investment decisions of households and firms will typically depend on the relative costs of financing the spending in question.

The composition of public debt can obviously be classified along various lines, each one raising its own particular policy issues. Government debt instruments thus differ with respect to tax treatment, the extent to which they are traded in well functioning second-hand markets, and whether they are subject to various regulative restrictions. However, the dimension most commonly associated with debt management policy is the maturity structure of government debt. In Table 1.1 we record the development of the average time to maturity of outstanding public

Does Debt Management Matter?

TABLE 1.1   *Average of remaining years to maturity of public debt outstanding, 1980–1987*

	USA	Japan	Germany	UK	Sweden
1980	3.7	5.2	4.5	12.2	4.4
1981	4.0	4.5	4.2	12.4	n.a.
1982	3.9	3.8	4.0	12.1	n.a.
1983	4.0	3.8	3.9	11.9	n.a.
1984	4.5	3.7	4.0	10.7	n.a.
1985	4.9	3.9	3.8	10.4	3.5
1986	5.2	4.1	4.1	10.4	3.6
1987	5.7	4.8	4.1	10.9	3.3

Source: central banks of the respective countries

debt for a few OECD countries.[3] We see that there is no uniform development: while the average time to maturity has fallen for some countries (e.g. UK and Sweden), it has remained fairly stable, or has even increased somewhat, for others (e.g. the USA).

The traditional aim of debt management has been to minimize the government's costs of borrowing. In a series of papers in the late 1950s and early 1960s, several researchers[4]—with James Tobin as the leading proponent—showed that debt management could also be used as an instrument for stabilization policy. The rapidly growing volumes of public debt have more recently led to a renewed academic interest in the economics of debt management. The insights in the older literature have thus been 'rediscovered' and have spurred a growing body of research.

The aim of the present paper is to provide a presentation and a critical discussion of this research. However, we have not aimed at writing a survey article in its ordinary sense—for that purpose, our discussion is far too selective—but rather at providing a review of some topics in the literature that we find particularly interesting and intriguing. In particular, it should be stressed at

[3] The data have been obtained from the central banks of the respective countries.

[4] See e.g. Rolph (1957), Musgrave (1959), Brownlee and Scott (1963), Okun (1963), and Tobin (1963).

the outset that the empirical analysis of the present paper is of an explicit partial equilibrium nature. A comprehensive discussion requires analysing not only the effects of debt management on relative asset yields, but also how these yield adjustments ultimately affect the macroeconomy.[5]

Chapter 2 examines some of the basic conceptual issues involved. A distinction is made between a 'pure' debt management policy and monetary and fiscal policy. We review some of the neutrality theorems of government finance and discuss under what set of conditions debt management operations are of no economic significance. We also digress on the potential targets and instruments of debt management, concluding that there is no simple single-dimensional characterization of government debt policy: any particular debt management operation involves choosing between a continuum of different 'debt attributes' to attain equally multi-dimensional goals.

In Chapter 3 we turn to the basic portfolio balance approach to debt management, which in one form or another constitutes the backbone of the empirical studies in the area. Chapter 4 deals with some of the problems involved when implementing the basic model. As the key asset substitutabilities and risk premiums governing the effects of debt management depend on the subjective risk perceptions of investors, we typically expect estimates of the effects of debt management to show little stability over time. After comparing the results obtained using time-series methods for inferring agents' risk perceptions, we find that this anticipation is indeed supported by our data. We also examine the robustness of results to the choice of data interval. After comparing the results obtained using quarterly and monthly data, we conclude that the frequently neglected problem of temporal aggregation bias should be of more concern in future work.

A problem associated with the vector autoregression approach to investors' risk perceptions used in Chapter 4 is that it is inherently backward looking. In any period, the elements of the return covariance matrix entering investors' asset demand functions are estimated using historical data. More realistically, investors use whatever information they have available when

---

[5] For an example of such a 'general equilibrium' treatment, see the contribution by Benjamin Friedman in Part II below.

Does Debt Management Matter? 11

assessing the riskiness of different assets. In addition to past return data, the information set may thus include 'noise' in the form of non-fundamental market rumours and 'news' concerning market fundamentals (e.g. changes in monetary and fiscal policy). Chapter 5 suggests a direct method to deal with these issues. Using options data and standard models of option pricing, we compare the 'implicit' standard deviations of underlying asset returns with those derived from a conventional time-series approach. As the 'options' variances in any given period mirror whatever information investors find relevant, we typically expect them to exhibit little resemblance to those implied by the vector autoregression procedures.

Chapter 6 turns to one of the analytical shortcuts underlying the basic 'work-horse' model used in previous chapters. Most of the literature examines the effects of debt management on expected asset returns while (implicity) taking current asset prices as given. As the return adjustments on most long-term assets occur via changes in current asset prices, this procedure is far from innocent. For instance, as current asset prices change, so does aggregate private wealth, which in turn implies wealth effects, which affect the pattern of equilibrium returns. In Chapter 6 we thus drop the standard assumption of exogenous current asset prices and examine the effects of debt management when *all* yield adjustments occur through changes in current prices. Chapter 7 concludes the paper by presenting some policy conclusions and some suggestions for future research.

2

Some General Concepts

2.1 What is Debt Management?

To study debt management, we first have to define it. The concept of government debt seems fairly straightforward: it should include all assets issued by the government. The most obvious examples of such assets are Treasury bills and long-term bonds, but governments have other liabilities too, some of which will be discussed below. Government debt instruments are held by households, by the corporate sector, and by financial institutions. Some of these assets are also held by the central bank, perhaps as a result of open-market operations. Here we will include the central bank in the debt-managing authorities and thus consider as government debt only those assets held by the private sector. There is some ambiguity as to which assets to include. Various kinds of government bonds should, of course, be included, as well as non-marketable securities like deposits of the private sector in the central bank or in other government agencies. Money is also an asset issued by the government and held by the private sector, and should consequently be regarded as government debt. At the very general level, the vast and vague array of government commitments to pay out money in the future in the form of social security benefits, veteran's pensions, etc., should also be included.

Public debt management can be defined as the government's (including the central bank's) choice regarding the composition of the outstanding stock of all the securities entering the liability side of its balance sheet. However, what should count is not the government's gross liabilities, but its net position *vis à vis* the private sector. This means that the government's position with regard to securities appearing on the asset side of its balance sheet, such as land, or stock in private companies, could also for all practical purposes be regarded as debt management: by

Does Debt Management Matter?　　　13

trading in such assets, the government might affect the equilibrium in asset markets and thereby the pattern of asset yields.

When we want to study debt management empirically, we have to exclude many of the kinds of actions that in principle ought to be considered as debt management. We will thus disregard all government trading in the markets for land and common stock. Also, the 'privatization' of government-owned companies, which has recently attracted so much attention (see e.g. Yarrow 1986), will not be treated here. This is mainly because the selling of government-owned firms in most countries still tends to be of a fairly minor magnitude. It does not therefore have any noticeable impact on financial markets, but seems to be motivated mainly by ideological considerations and by attempts to increase the productivity and efficiency of these firms, rather than by concern about the equilibrium configuration in asset markets.

The question of how to treat money is another difficult problem. In principle, money should be included in the concept of public debt, and there is no practical reason to exclude it because of a lack of data. On the other hand, there is a distinction between money and other assets with respect to the theoretical tools used to study them; the transactions approach typically used for studying monetary phenomena is quite different from the portfolio approach commonly used for studying the markets for other financial assets. And while there is a fairly well developed and widely accepted theory of the demand for bonds, no such generally accepted theory of the demand for money exists. Here we will therefore limit our attention to 'pure' debt management operations, which we regard as being separate from monetary policy. We will also try to treat debt management policy as separate from fiscal policy, although the distinction sometimes seems as difficult as that from monetary policy. Unless one possesses a full-scale model of the macroeconomy, within which all three types of policy can be studied simultaneously, specialization in one type of policy requires that it can be reasonably isolated from the others. We will therefore develop in some detail how such an isolation can be conceived.

In any time period, the government's spending can be financed in three ways: by taxation, by money creation (i.e. by the inflation tax), and by borrowing. The proper mix of these three

14   *Jonas Agell and Mats Persson*

ways of financing is an intricate but nevertheless well defined
optimization problem which was first formulated by Phelps
(1973). In this formulation, fiscal policy (i.e. the choice of the
amount of government spending and of tax financing), monetary
policy (i.e. the choice of the amount of money creation), and
government borrowing all have to be determined simultaneously
as integrated parts of 'the general macroeconomic problem of
public finance'.

Solving this general problem in practice would require a full-
scale model of the entire economy. On a less ambitious scale,
it is possible to study the effects of changing the composition
of government debt while taking fiscal and monetary policy as
given. In a given period t, we can write the government's budget
constraint as

$$G_t + rB_{t-1} = T_t + M_t - M_{t-1} + B_t - B_{t-1} \qquad (1)$$

where variables on the left-hand side refer to the government's
expenses (consumption G_t and interest on previous period's debt
rB_{t-1}) and variables on the right-hand side refer to the three
sources of revenue (tax revenues T_t, money creation $M_t - M_{t-1}$,
and net borrowing $B_t - B_{t-1}$). One would then like to identify
changes in G_t and/or T_t with fiscal policy, and changes in $(M_t -
M_{t-1})$ with monetary policy. For given values of G_t, T_t, and
$(M_t - M_{t-1})$, i.e. for a given value of net borrowing $(B_t - B_{t-1})$,
we say that there is scope for debt management if changes in
the composition of the debt B_t can affect the real side of the
economy.

Equation (1) is in many respects incomplete. For example, we
have not distinguished between debt instruments of different
maturities, with different tax treatments, etc., but have aggregated all interest rates into a single rate r and all instruments
into a single number B_t. Depending on the composition of the
aggregate B_t, a variety of future cash flows for the government
is conceivable. For example, if B_t consists only of short-term
debt, the cash flow at time $t + 1$ in the form of interest payments
and amortization will be large. On the other hand, if B_t consists only of long-term debt, the cash flow at time $t + 1$ and $t + 2$
will be relatively moderate; if all bonds are discount bonds,
there will not even be any interest payments until the date of

Does Debt Management Matter? 15

redemption, and the cash flow at, say, date $t + 10$ or $t + 20$ will be correspondingly large.

While equation (1) treats only what happens at time t, a full theory of debt management must thus also take into account the fact that different borrowing strategies will have different implications for the government's finances in the future. Since such intertemporal aspects have to be taken into account, it is difficult to keep debt management distinct from fiscal and/or monetary policy.

In the short run, things are simple. We can always conceive of a change in the composition of the aggregate B_t which does not affect its size, i.e. does not affect G_t, M_t, or T_t, at time t. This is the standard way of analysing debt management, and it is the approach taken here. In the longer run, however, such a change might have implications for the size of future values of total debt, implying that future taxes or expenditures may have to be changed. In principle, one would like to define long-run debt management in the following way. Consider given sequences $\{G_t\}_0^\infty$, $\{T_t\}_0^\infty$, and $\{M_t\}_0^\infty$. These sequences have to satisfy the government's solvency constraint, which means that the sum of the present discounted values of these expenditure and income streams has to be non-negative. Given such sequences, we can consider changes in the composition of the sequence $\{B_t\}_0^\infty$ that leave the entire sequences in G_t, T_t, and M_t unaffected. In that way, debt management will be distinct from fiscal and monetary policy even in the long run.

The question naturally arises whether such a strict definition of debt management is meaningful, i.e. whether not all changes in the composition of B_t will have to be countered by changes in some future G_{t+s}, T_{t+s}, or M_{t+s}. The answer is that this depends on the model used; at this general level, where we have not yet specified any model to which to apply our definition, we can easily conceive of an infinite number of changes in B_t that leave all future fiscal and monetary variables unaffected. For example, if the economy is such that the Ricardian equivalence theorems of debt management[1] hold, changes in the composition of public debt have no effects whatsoever on the economy, and thus an

---

[1] This is discussed in Sect. 2.2 below.

16   *Jonas Agell and Mats Persson*

infinite number of policies satisfying our definition is conceivable—whether or not any of these policies is warranted is another matter.[2] On the other hand, if the economy is of a Keynesian variety, where debt management does affect macroeconomic variables, future fiscal and monetary instruments generally have to adjust, thus rendering it impossible to isolate the effects of debt management *per se*.

Here, however, these explicitly dynamic aspects will be disregarded, and we will follow the standard approach to debt management by considering a simple, atemporal model of the economy. It turns out that many of the issues that make this approach problematic have to do with the difficulties of trying to mimic a multi-period reality by an atemporal model (see e.g. Chapter 6 below). Still, no practical alternative to this approach seems to exist as yet; existing intertemporal models of capital market equilibrium are still at a stage of development where empirical implementation for the purpose of analysing government debt management seems extremely difficult.[3]

The definition of debt management in an atemporal (or one-period) setting is now straightforward. Debt management means that the government changes the composition of outstanding debt by operations in the financial market, i.e. by selling as much (in value terms) of one type of instrument as it simultaneously buys of another. If sales were not matched by purchases of exactly the same value, the difference would have to be met by changes in the fiscal and/or monetary variables, and the analysis of such changes requires quite different tools.

From this definition it follows that (for example) open-market operations, where the central bank buys or sells bonds against money, have to be considered as a mixture of monetary policy (in the sense that they change the money stock) and debt manage-

---

[2] In the models of Lucas and Stokey (1983) and Persson *et al.* (1987), the government at time *t* can choose the optimal values of the fiscal and monetary variables, and then has enough degrees of freedom left to choose any maturity structure of the public debt. Thus, debt management (which in this case means choosing a particular maturity structure) does not effect the government's objective function, but could be used to achieve something else, for example to impose time consistency on future governments.

[3] For example, the recent theoretical work of Cox *et al.* (1985a), integrating intertemporal capital asset pricing models and linear stochastic production theory, still goes a long way from providing a truly *macroeconomic* framework for empirical policy analysis.

Does Debt Management Matter?   17

ment (in the sense that the government has to decide upon the composition of the bonds sold or purchased). To concentrate on the latter, therefore, we will consider the monetary parts of such mixed policies as given. It also follows that we have to disregard the implications of foreign borrowing. If the government changes the mixture of domestic and foreign borrowing, this will by definition affect the money stock; increased borrowing abroad and a corresponding reduction in the borrowing at home will increase the money stock, and vice versa. To concentrate on pure debt management, therefore, we will consider only domestic borrowing.

Our definition of debt management draws an equally sharp line at fiscal policy. By studying regular market operations involving, say, the sale of long-term bonds in exchange for short-term bonds, we limit ourselves to letting fiscal policy be a given parameter determined outside the analysis. There is, however, another approach to debt management analysis, emphasizing the effects of introducing new debt to finance an increase in government spending, which warrants a brief digression.

This latter approach—involving a mixture of debt management and fiscal policy—is intimately related to the crowding-in– crowding-out issue in macroeconomics. The question addressed is to what extent extra government spending financed by bonds 'crowds out' private investment and reduces the potency of fiscal policy. As shown by Tobin (1963) and Friedman (1978), the answer to this question is highly dependent on the assumptions made concerning the asset menu available to investors. In a portfolio balance model incorporating different types of bonds, Friedman showed how bond-financed deficits could actually lead to 'crowding-in'. The mechanism behind this counter-intuitive result is that, if the new government spending is financed by the issue of bonds that are distant, rather than close, substitutes to corporate equity, portfolio readjustments of investors will lead to increased demand for equity. Thus, equity returns go down and private investment increases.

The 'market operation' approach that is adopted in what follows obviously does not deal with the effects of an increase in government spending. Thus, the question of crowding-in or crowding-out in its ordinary sense is not relevant here. What matters is rather the portfolio crowding-out associated with regular market

18 *Jonas Agell and Mats Persson*

operations. The policy issues involved are thus somewhat less exciting when one limits oneself to this approach—or at least are somewhat further away from the question raised in the popular debate.

This narrow-mindedness can be defended on two grounds. First, to properly study the effects of mixed debt management–fiscal policy experiments would require a model of the entire macroeconomy. In the absence of a coherent view on a variety of theoretical and empirical issues, ranging from the signs and size of key elasticities to the proper integration of the monetary and real sectors of the economy, this is obviously too demanding a task.

For example, when examining the effects of new debt issues, the question of Ricardian equivalence becomes immediately relevant. If the government increases its borrowing, we have to consider whether or not the public regards these bonds as net wealth. In the debt management literature dealing with the effects of new debt issues rather than regular open-market operations, one typically assumes that all new government debt is regarded as net wealth by investors (see e.g. Friedman 1985; Frankel 1985). The plausibility of this assumption is of course questionable: it could well be argued that private agents, at least to some extent, take into account the future tax increases required to service the debt.[4] However, for the time being it seems appropriate to leave this particular form of Ricardian equivalence for the future research agenda.[5]

Second, examining the effects of debt management in the form of open-market operations is an important intermediate step when analysing the traditional crowding-out issues. Only by first understanding the effects of debt management *per se* can we hope to make progress in understanding its interaction with fiscal policy.

---

[4] See Werin (1990) for a model incorporating partial discounting of future tax payments.

[5] It should be stressed right away that we cannot as easily circumvent the Modigliani–Miller theorems of debt management which extend the basic Ricardian irrelevance principle to other financing decisions than the choice between debt and taxes. See the following section for further discussion.

Does Debt Management Matter?

2.2 When Debt Management does not Matter

A well-known result in macroeconomics is that under certain conditions the way the government chooses to finance its expenditures is immaterial to the real allocation in the economy. Financing by taxation or financing by borrowing will result in exactly the same budget sets for the agents, who will therefore follow the same intertemporal consumption plans. This is the so-called Ricardian 'equivalence theorem' (see Barro 1974).

In this section we will consider under what set of assumptions debt management operations (rather than the choice between borrowing and taxation) have no effects on real magnitudes. We will review some of the recent 'Modigliani–Miller' theorems of government finance, which, using arguments similar to those of Modigliani and Miller (1958), extend the basic Ricardian irrelevance principle to government financing decisions other than the choice between debt and taxes. In models where agents correctly take account of the intertemporal government budget constraint and the stochastic processes governing asset returns, irrelevance propositions have thus been derived concerning, among other things, index bonds, open-market operations involving money and bonds, and the maturity composition of government debt.

These theoretical developments are obviously in glaring contrast to most of the empirical work on debt management. In the latter literature—following Tobin (1963) and Friedman (1978)—the presumption is that debt management, at least in principle, ought to matter: by altering relative asset supplies, the government can affect relative asset yields and hence the rest of the economy. Failure to identify empirically significant effects of debt management on relative asset yields is typically interpreted as a sign of assets being close substitutes in investors' portfolios rather than as an indication of the empirical plausibility of various irrelevance theorems.

This discrepancy between recent theoretical developments and current empirical practice is obviously discomforting. Thus, depending on where one wants to lay the burden of evidence, the question seems to be either what confidence we can have in established empirical work with somewhat ambiguous theoretical

20   Jonas Agell and Mats Persson

foundations, or what credibility we can assign to theoretical work which is at odds with much of the empirical findings indicating the potential importance of debt management operations. In this section we will highlight some of the issues involved by examining the assumptions that must hold for neutrality of debt management to prevail. In particular, we will identify the kind of assumptions that we need to relax for debt management to regain its potency.

To pin-point the key considerations involved, we first need a bit of more formal analysis.[6] Assume an economy with a single good, which is consumed by a household sector and a government sector. Each household lives forever and supplies an exogenous amount of labour in each time period. The choice problem of the typical consumer consists of finding a sequence of asset holdings that maximizes the expected utility of lifetime consumption.

In every period t the government finances its expenditures by tax collections and by borrowing.[7] For notational simplicity, we assume that all government borrowing takes place in the form of discount bonds. At time t the outstanding quantity of government bonds maturing at time τ is $B(t, \tau)$. Each of these bonds promises its owner the receipt of one unit of the consumption good at τ. Denoting the current market price of a bond maturing at time τ by $p(t, \tau)$, the government budget constraint at time t is given as

$$G(t) + B(t-1, t) = T(t) + \sum_{\tau=t+1}^{\infty} p(t, \tau)[B(t, \tau) - B(t - 1, \tau)],$$

$$(2)$$

where all values are measured in terms of the underlying consumption good. The left-hand-side variables refer to the government's expenditures (consumption $G(t)$ and debt service) and the right-hand-side variables refer to the sources of revenue (tax revenue $T(t)$ and net borrowing).

Assuming also that all private securities are discount bonds promising to pay their owners one unit of the consumption good at the time of maturity, the ith household's budget constraint can be written as

---

[6] The following argument is due to Chan (1983).
[7] We abstract from financing by money creation.

Does Debt Management Matter? 21

$$X_i(t) + B_i(t - 1, t) + D_i(t - 1, t) - T_i(t)$$

$$= C_i(t) + \sum_{\tau=t+1}^{\infty} p(t, \tau)[B_i(t, \tau) - B_i(t - 1, \tau)]$$

$$+ \sum_{\tau=t+1}^{\infty} \pi(t, \tau)[D_i(t, \tau) - D_i(t - 1, \tau)]. \tag{3}$$

The left-hand side of (3) is disposable income after payment of taxes $T_i(t)$. The gross income consists of three parts: exogenous labour income $X_i(t)$; income from government bonds held by the household and maturing at time t, $B_i(t - 1, t)$; and income from private discount bonds held by the household and maturing at time t, $D_i(t - 1, t)$. Since we allow for short-selling of private assets, $D_i(t - 1, t)$ can be positive as well as negative.[8] The right-hand side of (3) defines the uses of disposable income: consumption $C_i(t)$ and savings, possibly negative, in government bonds and private securities.

The variable $\pi(t, \tau)$ is the market price at time t of a privately issued security maturing at time τ. In the absence of uncertainty, it must obviously be the case that $\pi(t, \tau) = p(t, \tau)$. In what follows, however, we will allow for uncertainty. Assuming that for every period t there are some states of the world where either the government or the private security issuers default on the bonds falling due (meaning that $B_i(\tau, \tau)$ or $D_i(\tau, \tau)$ are zero for some of the states realized at time τ), $\pi(t, \tau)$ and $p(t, \tau)$ are to be interpreted as the current market prices of risky consumption claims due at time τ. When the state-dependent pay-off structure on a government bond differs from the pay-off structure on a private bond of the same maturity, $\pi(t, \tau)$ will typically differ from $p(t, \tau)$. This corresponds to the case when government and private bonds are imperfect substitutes.

The polar case is when government and private bonds are perfect substitutes. Then the pay-off structure on a government bond is identical to the pay-off structure on a private bond of the same maturity. With competitive financial markets, $p(t, \tau)$ must then equal $\pi(t, \tau)$.

The existence of a perfect private substitute for public debt is

---

[8] For our purposes it suffices to note that the private bonds can be viewed either as claims against different households or as debt instruments issued by private firms.

22   *Jonas Agell and Mats Persson*

one of the key assumptions needed to establish the irrelevance of government debt management. That this must be the case is hardly surprising. Whenever government and private bonds are less than perfect substitutes—meaning that the pay-off structure on government bonds cannot be replicated by a portfolio of private bonds—and when there are also binding non-negativity constraints on private holdings of government bonds, government borrowing in general and debt management in particular will alter the risk–return opportunity set confronting investors. Hence market-clearing asset prices and resource allocation will be affected by government financial operations.

All taxes are assumed to be of the lump-sum variety. In every period t the household pays a known fraction θ_i of total government tax revenue $T(t)$. Thus, we have

$$T_i(t) = \theta_i T(t), \tag{4}$$

where $\Sigma_i \theta_i = 1$.

We next assume that we are in an initial equilibrium at time t_0 where all asset markets clear at prices $p^*(t_0, \tau)$ and $\pi^*(t_0, \tau)$. This equilibrium is consistent with each consumer having chosen optimal consumption–portfolio strategies given the household budget constraint (3), the government budget constraint (2), and the tax-sharing rule (4).

One of the key issues in the economics of government debt management is whether changes in the maturity composition of government debt affect the economy. This question is easily addressed using the simple framework just presented.[9] As a matter of definition, we first say that debt management is irrelevant if a change in the maturity composition of government debt at time t_0 leaves the sequences of equilibrium prices $\{p^*(t, \tau)\}_{t=t_0}^{\infty}$ and $\{\pi^*(t, \tau)\}_{t=t_0}^{\infty}$ unchanged.

Assume that the government at time t_0 sells one unit of long-term bonds at a price $p(t_0, t_L)$ and uses the proceeds to buy $p(t_0, t_L)/p(t_0, t_S)$ units of short-term bonds $(t_L > t_S)$. This financial perturbation policy represents a 'pure' debt management operation in the static sense defined in the previous section; for, given values of $G(t_0)$ and $T(t_0)$, the government changes its borrowing

---

[9] See Chan (1983) and Stiglitz (1983) for derivations of irrelevance propositions concerning the maturity structure of government debt.

Does Debt Management Matter?

mixture at time t_0 in a manner compatible with the financing constraint (2). This financing constraint must, however, also hold at time t_S and t_L. At time t_S aggregate tax revenue $T(t_S)$ is reduced by $p(t_0, t_S)/p(t_0, t_S)$, reflecting the reduced amount of government debt coming due. At time t_L taxes are raised by one unit in order to finance the amortization of the long-term bonds issued at time t_0.

One possible portfolio response of households to the financial perturbation policy is as follows. At time t_0 the ith household buys a fraction θ_i of the newly issued long-term government bonds and finances this by selling $\theta_i p(t_0, t_L)/p(t_0, t_S)$ units of short-term government debt. As this operation simply involves adjusting the maturity composition of a given wealth portfolio, it leaves feasible consumption at time t_0 unchanged.

It is also easily established that this particular portfolio strategy leaves feasible consumption at time t_S and t_L unchanged. At time t_S the household experiences a reduction in state-dependent capital income corresponding to the quantity $\theta_i p(t_0, t_L)/p(t_0, t_S)$ of short-term debt sold at time t_0. This income loss is, however, balanced by a state-dependent tax cut of the same magnitude; as aggregate taxes are cut by an amount reflecting the reduced quantity $p(t_0, t_L)/p(t_0, t_S)$ of short-term government debt coming due at time t_S, the tax-sharing rule (4) implies that household-specific taxes are cut by an amount corresponding to the quantity $\theta_i p(t_0, t_L)/p(t_0, t_S)$ of short-term government debt sold by the household at time t_0. Similarly, at time t_L the household obtains additional state-dependent capital income, corresponding to the quantity θ_i of long-term government debt bought at time t_0, which is exactly matched by a state-dependent tax hike used to finance the increased quantity of government debt coming due at t_L.

We have thus shown that for each household there exists a particular portfolio strategy which implies that debt management operations of the government have no effects on the intertemporal opportunity sets of households. It is also immediately clear that this portfolio strategy satisfies our irrelevance condition of government debt management, meaning that changes in the maturity composition of government debt leave the sequences $\{p^*(t, \tau)\}_{t=t_0}^{\infty}$ and $\{\pi^*(t, \tau)\}_{t=t_0}^{\infty}$ of equilibrium asset prices unchanged. At time t_0 the supply of long-term government debt increases by one unit and the supply of short-term government debt decreases by $p(t_0, t_L)/p(t_0, t_S)$ units. Asset market equilibrium at time t_0 is, however, unaffected: aggregate household demand for long-term government bonds increases by $\Sigma\theta_i = 1$ unit and aggregate household demand for short-term government bonds decreases by $\Sigma\theta_i p(t_0, t_L)/p(t_0, t_S) = p(t_0, t_L)/p(t_0, t_S)$ units. Also, since household budget constraints are unchanged at time t_S and t_L, it is trivially true that asset markets clear at the old set of equilibrium prices at t_S and t_L. Consequently, the maturity structure of government debt is immaterial to the price sequences $\{p^*(t, \tau)\}_{t=t_0}^{\infty}$ and $\{\pi^*(t, \tau)\}_{t=t_0}^{\infty}$ and to the real allocation in the economy.

The next important question is under what set of assumptions the presumed household portfolio strategy, implying neutrality of government debt management, will actually be the one chosen by households maximizing expected utility. This question is obviously crucial. If there is no optimizing portfolio strategy corresponding to the one postulated above, the irrelevance proposition of government debt management is devoid of economic content.

Irrespective of the form of household utility functions, the following assumptions are sufficient for optimizing households to actually choose the postulated portfolio strategy. First and foremost is the assumption that price-taking households do 'pierce the government veil'; households correctly infer the interrelations between private and government budget constraints and fully realize any change in future state-dependent tax liabilities, induced by current-period debt management. Second, all taxes must be lump-sum, thus excluding deadweight losses related to the consumption–saving decision of households. Third, the tax-sharing rule (4) does not change over time, implying the absence of redistributional effects. Fourth, there must be a perfect private substitute for every type of government bond, meaning that government debt policy provides no new trading opportunities (see e.g. Chan 1983; Stiglitz 1983).

When asset prices remain unchanged, these assumptions ensure that the intertemporal budget constraints of households are unaffected by government debt management operations. As a consequence, the typical household cannot increase welfare by changing its original intertemporal consumption plan in response

Does Debt Management Matter? 25

to the financial perturbation policy of the government. But the household portfolio strategy consistent with both unchanged consumption and equilibrium in asset markets at the old set of asset prices is the one already suggested. By buying and selling government bonds in proportion to its fixed tax share θ_i, the typical household thus maintains its original welfare.

The Modigliani–Miller flavour of the irrelevance argument is obvious. In the Modigliani–Miller economy rational investors, seeing through the corporate veil and operating in perfect capital markets, undo the effects of corporate financial decisions by adjusting their portfolios. In the economy underlying the irrelevance propositions of government debt management, investors pierce the government veil and reverse the effects of government financial decisions by appropriate adjustments of private portfolios.

In this section we have seen how invoking conventional Modigliani–Miller assumptions renders changes in the maturity composition of government debt immaterial to the real economy. The same type of reasoning can be applied to other debt management operations. Parallel irrelevance propositions have thus been derived concerning open-market operations involving money and indexed bonds (Peled 1985) and money and real assets (Chamley and Polemarchakis 1984; Wallace 1981).

What, then, are the main insights to be gained from the Ricardian debt management results? First, casual empiricism suggests that their underlying assumptions are far from descriptive. In the real world, capital markets are less than perfect; taxes are redistributive and distortionary; empirical evidence indicates that agents do not fully discount future tax payments;[10] at least some government debt instruments (e.g. tax exempt government bonds) have no private counterparts; etc. In most situations relevant to the real world, we would thus expect debt management to affect the real economy. At the most practical level, we may therefore safely conclude that the Modigliani–Miller analysis of debt management is of little direct significance for the day-to-day business of the authorities managing the government debt.

However, viewed as an exercise in modelling design, the

---

[10] See Bernheim (1987) for a review of the evidence.

26   *Jonas Agell and Mats Persson*

Ricardian irrelevance literature raises a crucial empirical issue. By constructing models of hypothetical economies, incorporating carefully worked out intertemporal budget constraints, specifications of underlying return-generating processes, and behavioural functions consistent with basic microeconomic principles, this literature indicates the need to specify *structural* models of the economy. This argument gains particular force in an empirical context. Quantitative evaluations of the effects of debt management using various 'reduced-form' models of portfolio choice and asset markets are thus susceptible to the well-known Lucas (1976) critique of econometric policy evaluation: when one cannot econometrically identify 'deep' structural parameters relating to preferences and basic asset technology, the estimated response of asset returns to a shift in government borrowing will not be policy-invariant.

In view of the current state of financial economics, the idea of estimating truly structural models—which among other things consistently integrate real and financial markets—seems very difficult to honour in practice. It does, however, serve as a necessary reminder of some of the shortcomings of the traditional Tobin-type portfolio balance approach to debt management adopted in subsequent sections. By suppressing intertemporal linkages between private and government budget constraints, and by taking observed asset return series as a given datum rather than as the outcome of some underlying stochastic return-generating processes, this pragmatic approach obviously represents a second-best solution to empirical debt management analysis. Thus, the Lucas critique applies and all conclusions must be carefully guarded.

2.3 Objectives and Instruments of Government Debt Management

To the extent that changes in the composition of government debt do affect the real economy, we may ask to which objective one should assign debt management policy. At a very general level, the answer is straightforward. Debt management should be combined with other monetary and fiscal policy instruments so as to maximize a properly defined social welfare function. At a

Does Debt Management Matter?   27

somewhat lower level of abstraction, the issue becomes more involved. In the real world the government welfare function is thus a rather vacuous concept; typically, policy instruments are assigned to various intermediate targets which only in a very loose sense relate to some underlying basic objective function. The literature following Tobin (1963) has typically treated government debt policy as a component part of stabilization policies aimed at controlling aggregate demand. Thus, the effectiveness of debt management policy is typically discussed only in terms of the ability to influence corporate investment activity by controlling Tobin's q of the corporate sector. This is obviously a rather narrow view of the scope of debt management—we can easily list some other goals where government debt policy may make a difference.

First, the size of government interest payments is without economic significance whenever the government can use lump-sum taxes to finance current expenditures. In the real world, however, all tax instruments (including the inflation tax on the real cash balances of investors) have negative side-effects on resource allocation and economic efficiency. Recent empirical work indicates that these effects may be substantial (see e.g. Jorgenson and Yun 1986). Consequently, the size of the government's interest bill has real consequences: each dollar's worth of reduction in interest cost implies that less revenue has to be raised through distortionary taxes. This suggests that a second, particularly simple, policy rule for the national debt office would be to promote allocational efficiency by choosing a pattern of government borrowing which minimizes interest cost. This is in fact the traditional, pre-Tobin, goal of debt management.

Second, to the extent that the pattern of asset holdings differs across households, any change in the structure of asset yields (whether caused by debt management policy or by some other exogenous disturbance) will have distributional implications. Since existing data typically show that the structure of portfolio holdings differs widely across different types of investors, (King and Leape 1984), we cannot in any way exclude the possibility that debt management policy may have first-order effects on the distributional goals of the government.

Third, in the menu of assets facing the investors, foreign assets may play a large role. Changes in the composition of public debt,

28   *Jonas Agell and Mats Persson*

affecting the equilibrium asset yields, also affect the relative
yields of foreign and domestic assets. This gives rise to changes
in the capital account (if the exchange rate is fixed) or in the
exchange rate (if it is flexible), which in turn has consequences
for various macroeconomic variables.

To which, if any, of these goals should we assign debt management policy? An important insight in the literature on optimal
policy design is that very restrictive assumptions are required
before we can assign any particular policy instrument to one
particular policy objective. In the general case, with at least as
many objectives as instruments, the optimal use of different
instruments will depend on all the economic objectives of the
government. This also implies that it is less meaningful to analyse
the effects of debt management in terms of only one single policy
objective, like controlling q or minimizing government borrowing
cost. In the special case of a small open economy, where all asset
yields are internationally determined, the only conceivable goal
of debt management is to minimize government borrowing costs.
However, in the case where at least some asset yields depend on
relative domestic asset supplies, optimal policy design requires
that debt management is combined with other policy instruments
in such a manner that some overall government loss function is
minimized.

We have so far discussed debt management policy as if it were
a homogeneous and clearly defined concept. This is, however,
an oversimplification. As already noted in the Introduction, in
formulating its debt management policy the government has a
wide variety of choices concerning the type of debt instrument
used and the extent to which regulative elements are warranted.

A first important distinction is that between market-oriented
and regulative debt management policies. A regulative debt
policy works through the use of interest ceilings and reserve
requirements forcing private investors to acquire government
debt instruments at interest rates below those that would exist in
competitive financial markets. A market-oriented debt policy
presumes reciprocity between lenders and borrowers. Here,
the government must adjust its borrowing conditions in such a
manner that investors willingly hold the outstanding volume of
government debt instruments.

Which of these debt management regimes is preferable? In

Does Debt Management Matter?

principle, the advantage of a regulative debt policy is that the government, by defining its own advantageous borrowing conditions, can reduce the need for alternative financing through distortionary tax instruments. The main disadvantage is, of course, that a regulated pattern of interest rates provides misleading information of the social costs involved when evaluating investment projects, which in turn leads to an inefficient allocation of capital across sectors. We cannot a priori decide which of these effects is the dominating one. Less surprisingly, the choice between a regulative and a market-oriented debt management policy must therefore be based on empirical considerations, which are far beyond the scope of this paper.[11] Consequently, the fact that, in what follows, we lay stress upon various aspects of a market-oriented debt management policy should not be interpreted as the outcome of an implicit cost–benefit analysis of the efficiency of a regulative debt policy. Instead, it is to be seen as a reflection of the rapid institutional developments in financial markets in the last decade, which make it likely that government debt policy henceforth will be shaped within a broader context of market-oriented credit policies.

The government has also, within the confines of a market-oriented debt management policy, a great deal of latitude concerning its choice of borrowing instruments. Thus, in the real world government debt can be classified along a number of different—but not mutually exclusive—dimensions, each one having its own particular implications for the performance of debt management policy. The first one is simply the time to maturity at date of issue. In most of the literature, this is considered the central choice variable of debt management policy: the task of the national debt office is to choose a maturity composition of government debt which is optimal in the sense of achieving a certain impact on the relevant economic target variables. The main channel for this kind of maturity-oriented debt policy is the yield curve. From a policy perspective, the usefulness of changes in debt composition along the maturity dimension is therefore highly dependent on the ability of the debt office to affect the yield curve in a systematic and predictable manner.

A second distinction is whether government borrowing in-

[11] See e.g. Tobin (1963) for a general discussion.

30 _Jonas Agell and Mats Persson_

struments are fully or partially exempted from taxes. Thus, in many countries governments issue both tax-exempt and taxable debt. For instance, in the USA a large fraction of the debt issued by state and local governments is tax-exempt; in Sweden, debt instruments intended mainly for household investors are given a tax-sheltered treatment. There are no clear-cut empirical or theoretical conclusions concerning the implications for asset markets and the portfolio behaviour of investors of the government's choice between borrowing in tax-exempt and taxable forms. In general terms, the effects, if any, will depend on the institutional characteristics of the economy, such as the overall structure of the tax system and the precise form of any constraints on borrowing and short sales facing private investors (see e.g. Auerbach and King 1983; McDonald 1983).

A third distinction is that between issues of marketable and non-marketable debt. While a marketable debt instrument may be freely traded among investors, a non-marketable debt instrument is a personal and non-transferable contract between the government and a particular lender. It should be noted that this distinction is independent of the distinction between regulatory and market-oriented borrowing as discussed above. Regulatory borrowing could well be performed by means of marketable debt instruments; in that case the investors are forced to buy newly issued bonds at an interest rate below the market rate, but as soon as they have bought them they are free to sell them in the market at a corresponding capital loss.[12] As argued in Musgrave (1959), issuing non-marketable rather than marketable debt makes it possible for the government to earn the profits of a discriminating monopolist: whenever the financial system is segmented, with different types of investors operating in different markets, the government can reduce its overall interest bill by offering different returns to different investors.

A final dimension of debt management policy is whether debt instruments are denominated in real or nominal terms. With very few exceptions,[13] existing debt instruments are defined

---

[12] This was the way, e.g., the Swedish life insurance companies were regulated between 1984 and 1987.

[13] Historically, governments in high-inflation countries have issued indexed bonds. Among the more recent examples are the UK, Argentina, Brazil, Chile, and Israel.

nominally, thereby exposing lenders to uncertainty concerning the future purchasing power of their debt holdings. To the extent that market failures prevent private agents from issuing indexed bonds, government debt instruments defined in real terms might benefit financial markets by offering an investment opportunity that would otherwise not exist.[14] Also, if index bonds are closer substitutes for equity than nominally defined government debt, debt management operations involving index bonds would increase the possibility of controlling q and corporate investment activity (e.g. Tobin 1963).

In sum, the above considerations emphasize that there is no simple single-dimensional characterization of debt management policy. In the real world, the design of debt policy involves choosing between a continuum of different packages of 'debt attributes' to attain equally multi-dimensional and possibly mutually inconsistent goals. As a consequence, any description of a particular debt policy in terms of only one variable, such as the maturity composition of government debt, conveys only partial information concerning its economic effects.

---

[14] Fischer (1983) gives a discussion of the advantages and disadvantages of index bonds.

3

The Portfolio Balance Approach to Debt Management

This chapter turns to the question of the extent to which changes in the composition of government debt can systematically shift the structure of asset yields. As emphasized in Chapter 2, this is in reality a multi-dimensional problem involving choices between debt instruments with different tax treatment, varying degrees of marketability, and different times to maturity. We will however in what follows limit ourselves to considering the effects of changing the maturity composition of government debt. Since the work of Rolph (1957), Musgrave (1959), Brownlee and Scott (1963), Okun (1963) and Tobin (1963), and more recently Friedman (1978), this is considered *the* problem of debt management. As the spending decisions of firms and households typically depend on the relative costs of financing, steering relative asset yields by altering the maturity structure of government debt is viewed as a potentially important policy target.

Since the theoretical work of Brownlee and Scott (1963) and Roley (1979), the effect of debt management has been analysed using the mean–variance approach to portfolio choice originating in the work of Markowitz (1959) and Tobin (1958). By incorporating this basic optimizing theory into a Brainard–Tobin capital accounting framework (see Brainard and Tobin 1968; Tobin 1969), one thus obtains a model linking asset supplies to the structure of relative asset yields. This basic 'work-horse' model has served as a vehicle for a number of studies trying empirically to isolate the effects of government debt management on relative asset yields. This chapter presents a simple reference model which illustrates the basic mechanisms involved.

We consider a representative investor who can invest his wealth W in n assets. In the empirical applications below we will set n = 3 and we will identify asset no. 1 with corporate shares, asset no. 2 with long-term bonds, and asset no. 3 with short-term

Does Debt Management Matter? 33

bonds. For the time being, the names of the assets are of no importance.

The investor is concerned about the *real yield* of his portfolio. Since there is always some inflation risk, all assets will thus be risky—even short-term bonds. In general, we may think of the investor as choosing those optimal portfolio and consumption rules that maximize a properly defined intertemporal utility function. If we further assume that the investment opportunity set changes in either a completely random or a non-stochastic manner over time (cf. Merton 1982), the intertemporal optimization problem can be represented by a sequence of single-period portfolio models. Denote the end-of-period wealth by $\widetilde{W}$. The investor chooses to invest a fraction α_i of his initial wealth W in asset i, where $i = 1, \ldots, n$, so as to maximize expected utility of end-of-period wealth. He thus solves the problem

$$\max_{\alpha_1, \ldots, \alpha_n} \quad E[U(\widetilde{W})] \equiv E[U(W(1 + r))]$$

$$\text{subject to } r_W = \alpha_1 r_1 + \alpha_2 r_2 + \ldots + \alpha_n r_n$$

$$\sum \alpha_i = 1,$$

where r_W is the (random) yield on the investor's portfolio, and r_i is the individual (random) yield on asset i.

When analysing this problem, it is common to assume that the utility function displays constant relative risk aversion, i.e. that $-xU''(x)/U'(x) = c$ for all x, where c is a positive constant.[1] Making this assumption, it is a standard exercise in financial economics (e.g. Friedman and Roley 1987) to show that the demand for assets is given by the system

$$\alpha = \frac{1}{c} Br^e + \Pi. \tag{5}$$

Here α is the n-dimensional vector of asset demands in terms of portfolio shares, i.e. $\alpha \equiv (\alpha_1, \alpha_2, \ldots, \alpha_n)'$, where α_i is the share of total wealth that the investor will invest in asset i. Further, r^e is the vector of expected real returns on the assets, $r^e \equiv (r_1^e, r_2^e, \ldots, r_n^e)'$, where $r_i^e \equiv E(r_i)$. The $n \times n$ matrix B contains

---

[1] For empirical evidence that this is a reasonable assumption, see Friend and Blume (1975).

34   *Jonas Agell and Mats Persson*

information about the variance–covariance properties of the assets and is given by

$$B \equiv [\Omega^{-1} - (1'\Omega^{-1}1)^{-1}\Omega^{-1}1 \, 1'\Omega^{-1}]$$

where Ω is the covariance matrix of the asset returns; i.e., the typical element of Ω is $\sigma_{ij} \equiv \mathrm{cov}(r_i, r_j)$. The n-dimensional vector 1 is the unit vector $(1, 1, \ldots, 1)'$, while the n-dimensional vector Π is given by[2]

$$\Pi \equiv (1'\Omega^{-1}1)^{-1}\Omega^{-1}1.$$

Equation (5) gives us asset demand in terms of portfolio shares. If we want to express it in value terms instead, we simply multiply (5) by the scalar W:

$$d = \left(\frac{1}{c}Br^e + \Pi\right)W, \tag{6}$$

where the elements d_i of the vector d tells us how many dollars the agent will invest in asset i.

Let us for expository reasons take a look at the three-asset case, and let us denote the individual elements in the B matrix and the Π vector by

$$B \equiv \begin{bmatrix} b_{11} & b_{12} & b_{13} \\ b_{21} & b_{22} & b_{23} \\ b_{31} & b_{32} & b_{33} \end{bmatrix} \text{ and } \Pi \equiv \begin{bmatrix} \Pi_1 \\ \Pi_2 \\ \Pi_3 \end{bmatrix}.$$

Since we know that the individual's budget constraint implies $\alpha_1 + \alpha_2 + \alpha_3 = 1$, the system (5) is actually linearly dependent. We can therefore drop one of the demand equations, for example the last one, obtaining

---

[2] A special case occurs when one of the assets, say the nth one, is risk-free, i.e. when $\sigma_{in} = 0$ for all i. Demand for the $n - 1$ risky assets then simplifies to $\alpha = (1/c)\Omega^{-1}(r^e - r)$, where α now is an $(n - 1)$-dimensional vector of wealth shares, r is an $(n - 1)$-dimensional vector with all elements equal to the (risk-free) return on asset n, r^e is the vector of expected returns on the $n - 1$ risky assets, and Ω is the $(n - 1) \times (n - 1)$ covariance matrix of the risky asset returns. The demand for the risk-free asset is then given residually as

$$\alpha_n = 1 - \sum_{i=1}^{n-1} \alpha_i.$$

Does Debt Management Matter? 35

$$\begin{bmatrix} \alpha_1 \\ \alpha_2 \end{bmatrix} = \frac{1}{c} \begin{bmatrix} b_{11} & b_{12} & b_{13} \\ b_{21} & b_{22} & b_{23} \end{bmatrix} \begin{bmatrix} r_1^e \\ r_2^e \\ r_3^e \end{bmatrix} + \begin{bmatrix} \Pi_1 \\ \Pi_2 \end{bmatrix}. \tag{5'}$$

Denoting the (exogenous) supply of assets by $(\alpha_1^s, \alpha_2^s, \alpha_3^s)'$, we can use (5') to obtain the market equilibrium conditions

$$\begin{bmatrix} \alpha_1^s \\ \alpha_2^s \end{bmatrix} = \frac{1}{c} \begin{bmatrix} b_{11} & b_{12} \\ b_{21} & b_{22} \end{bmatrix} \begin{bmatrix} r_1^e \\ r_2^e \end{bmatrix} + \begin{bmatrix} \Pi_1 + \dfrac{r_3^e}{c} b_{13} \\ \Pi_2 + \dfrac{r_3^e}{c} b_{23} \end{bmatrix}. \tag{7}$$

Since this system contains three unknowns (r_1^e, r_2^e, r_3^e) and two equations, only two of the unknowns can be determined. Treating r_3^e as exogenous (which has been indicated in (7) by grouping the terms containing r_3^e together with the parameters on the right-hand side) we can write (7) as

$$\alpha^s = \frac{1}{c}(br + k) \tag{8}$$

where α^s, r, and k are two-dimensional vectors and b is a 2×2 matrix consisting of the first two rows and columns of the three-dimensional matrix B in (5) above. Solving for the endogenous r, we have

$$r = b^{-1}(c\alpha^s - k). \tag{9}$$

This equation gives us the equilibrium asset yields r_1^e and r_2^e as functions of the exogenous supplies α_1^s and α_2^s, the coefficient of relative risk aversion, and the elements of the covariance matrix Ω.

Denote the elements of b^{-1} by β_{ij}. Thus,

$$b^{-1} \equiv \begin{bmatrix} \beta_{11} & \beta_{12} \\ \beta_{21} & \beta_{22} \end{bmatrix}$$

and we see from (9) that

$$\frac{\partial r_i^e}{\partial \alpha_j^s} = c\beta_{ij}. \tag{10}$$

36   *Jonas Agell and Mats Persson*

Equation (10) does not have any immediately intuitive interpretation, since the coefficient β_{ij} is a rather complicated function of the elements in the covariance matrix. However, for the special case when short-term debt is riskless ($\sigma_{i3} = 0$, $i = 1, 2, 3$), (10) reduces to

$$\frac{\partial r_i^e}{\partial \alpha_j^s} = c\sigma_{ij}, \qquad i, j = 1, 2. \tag{10'}$$

Here σ_{ij} is the covariance between the two risky assets; a positive covariance is thus equivalent to a positive derivative. For example, if the ith asset is corporate equity and the jth is long-term bonds, and if these assets are positively correlated, we see that lengthening the maturity structure of public debt increases the equity return and thereby 'crowds out' private investment.

Expression (10) thus gives us the effects on asset yields of government debt management. Note that this is debt management according to the definition in Section 2.1 above; the derivative $\partial r_1^e/\partial \alpha_2^s$ shows how the return on asset 1 changes if the supply of asset 2 is changed at the same time as α_1^s remains constant and α_3^s is changed so as to leave $\alpha_1^s + \alpha_2^s + \alpha_3^s = 1$.

Finally, a few words about taxes. What matters to the investor is of course the after-tax yield, and thus the expected return vector r^e and the covariance matrix Ω should be interpreted as referring to after-tax yields. Now, there is a simple way to incorporate this into the analysis. Denote by Ω and r^e the covariance matrix and the expected return vector of gross (before-tax) yields, and by $\hat{\Omega}$ and $\hat{r}^e$ the corresponding matrix and vector of net (after-tax) yields. Let t_i indicate the tax rate applied to asset i. Then the typical element in the $\hat{\Omega}$ matrix is

$$\hat{\sigma}_{ij} \equiv \text{cov} \left[(1 - t_i)r_i, (1 - t_j)r_j \right] \equiv (1 - t_i) (1 - t_j) \text{cov}(r_i, r_j)$$
$$\equiv (1 - t_i) (1 - t_j) \sigma_{ij},$$

where σ_{ij} is the element in the matrix Ω. Thus $\hat{\Omega}$ can be simply expressed in terms of Ω as

$$\hat{\Omega} = T\Omega T,$$

where T is an $n \times n$ diagonal matrix with elements $(1 - t_i)$ in the diagonal and zeros elsewhere. Similarly, we have

$$\hat{r}^e = Tr^e.$$

From now on, we will assume that tax rates are the same for all assets, i.e. that $t_i = t_j = t$ for all i, j. Then we obtain

$$\hat{\Omega} = (1 - t)^2 \Omega \quad \text{and} \quad \hat{\Omega}^{-1} = \frac{1}{(1 - t)^2} \Omega^{-1}.$$

Similarly, the return vector becomes

$$\hat{r}^e = (1 - t)r^e.$$

Inserting this into the demand system (5) yields

$$\alpha = \frac{1}{c} \frac{1}{(1 - t)} B r^e + \Pi,$$

where B and Π are defined as before in terms of the covariance matrix Ω of gross returns. Thus, the demand system will look exactly as it did without taxes, except for the fact that the degree of risk aversion has been multiplied by $(1 - t)$. Unless one has exogenous information on one of them, it will therefore be impossible to identify c and t separately.

From now on we will use the formulation (5), or its corresponding expression (6), for demand in value terms, but bear in mind that c represents risk aversion inclusive of the tax rate. In the empirical analysis below we will assume that c is equal to 4, a value which seems like a compromise between various estimates of relative risk aversion[3] and a reasonably realistic tax rate for the average investor. If however some reader has a strong opinion about the correct value of c, note that c occurs in the policy derivative (10) only as a scale factor. In the following chapters, where we compute numerical values of $\partial r_i^e / \partial \alpha_j^s$, readers who do not agree with our assumption of $c = 4$ can just shift the curves up or down, applying any other scale factor.

---

[3] The numerical value of Arrow–Pratt's measure of relative aversion is an unsettled empirical question. The analysis in Pindyck (1984) suggests a value for c of around 5 or 6; indirect evidence in Grossman and Shiller (1981) indicates a value in the neighourhood of 4; whereas the cross-sectional household estimates in Friend and Blume (1975) imply a value of at least 2.

4

Implementing the Basic Model by Using Historical Data

The basic empirical question involved when assessing the potency of government debt management is to what extent different assets in investors' portfolios are close substitutes. Several different approaches are used in the literature to estimate the relevant asset demand parameters, not all of which impose the constraints of mean–variance optimization discussed above.[1] In the following, however, we will make the admittedly strong assumption of taking the mean–variance model as a given datum. In a mean–variance context, the substitutability of different assets ultimately depends on the elements of the covariance matrix Ω. Rather than test the basic mean–variance model, the purpose of this chapter is to discuss different procedures for estimating the covariance elements to be fed into the model. Since the covariance matrix typically depends on the subjective risk perceptions of investors as well as underlying 'objective' return probabilities, estimating its elements is a far from trivial task.

We will start in Section 4.1 by using a very simple estimate of historical covariances. In Sections 4.2 and 4.3 we will make the analysis somewhat more sophisticated, employing the vector autoregression methods employed by Friedman (1985b, 1986) to estimate covariance structures. In particular, we will examine the stability over time of the implied debt management derivatives calculated according to (10) above. From a policy point of view, the stability issue is obviously crucial, as successful debt management rules require systematic and stable policy responses. The stability issue is further analysed in Section 4.4, where we examine the robustness of results to the choice of data interval. After comparing the results obtained using quarterly and monthly data, we conclude that the frequently neglected problem of

---

[1] See Frankel (1985) for an overview of the literature.

Does Debt Management Matter? 39

temporal aggregation bias should be of more concern in future work.

4.1 A Simple Estimate of the Covariance Matrix

The total yield of any asset consists of a dividend yield (d) and capital gain or loss (g), adjusted for inflation ($\dot{p}$):

$$\text{Total real yield} = d + g - \dot{p}.$$

Of these three components, the dividend can be regarded as certain[2] and the main uncertainty is therefore associated with the capital gain (or loss) and the rate of inflation. Thus, d should be included in the r^c terms of the previous chapter, but not in the covariance matrix Ω.

For the sake of illustration, we have computed the covariance matrix of the real capital gain ($g - \dot{p}$) for three types of assets— corporate stock, long-term government bonds, and three-month Treasury bills—using US quarterly data from 1960(1) to 1988(2) (IMF 1961–1988). When calculating the capital gains of the long-term bonds from the published interest rate series, we treated them as consols, which seems like a reasonable approximation. The (nominal) capital gain on stock is simply the change in the Standard & Poor's 500 share price index. The Treasury bills display no capital gains or losses: for them $g = 0$ in all periods, and the only risk is the one associated with $\dot{p}$.

The covariance matrix of these three yield series is given in Table 4.1(a). To facilitate comparison, the coefficients of correlation[3] are given in Table 4.1(b). We see first that the variance of the real yield on short-term bonds is very low; since the only source of uncertainty here is the inflation rate, which in turn varies very little on a quarterly basis, these bonds could almost be regarded as a safe asset. Second, we see that all three assets are positively correlated.

If instead we compute the correlations of the nominal yields, this pattern is changed. This is shown in Table 4.2(a). Here we

---

[2] We disregard the possibility of bankruptcy.

[3] The covariances σ_{ij} in Table 4.1(a) are related to the coefficients of correlation ρ_{ij} in Table 4.1(b) according to $\rho_{ij} = \sigma_{ij}/\sqrt{\sigma_{ii}\sigma_{jj}}$

40   *Jonas Agell and Mats Persson*

TABLE 4.1 *Covariations between the quarterly real yields of (1) corporate equity, (2) long-term government bonds, and (3) short-term government bonds: US data, 1960(1)–1988(2)*

		Corporate stock	Long-term bonds	Short-term bonds
(a)	*Covariance matrix*			
	Corporate stock	0.00457	0.00127	0.00023
	Long-term bonds		0.00401	0.00024
	Short-term bonds			0.00008
(b)	*Coefficients of correlation*			
	Corporate stock	1	0.30	0.38
	Long-term bonds		1	0.43
	Short-term bonds			1

Source: IMF, *International Financial Statistics*

TABLE 4.2 *Covariations between the quarterly nominal yields of (1) corporate equity, (2) long-term government bonds, and (3) the consumer price index: US data, 1960(1)–1988(2)*

		Corporate stock	Long-term bonds	CPI
(a)	*Covariance matrix*			
	Corporate stock	0.00418	0.00088	−0.00015
	Long-term bonds		0.00361	−0.00016
	CPI			0.00008
(b)	*Coefficients of correlation*			
	Corporate stock	1	0.23	−0.26
	Long-term bonds		1	−0.31
	CPI			1

Source: IMF, *International Financial Statistics*

have excluded the short-term bonds, since they do not have any nominal risk. Instead we have included a real asset, say land or consumer durables, and assumed that the price of that asset behaves according to the consumer price index. In Table 4.2(*a*)

Does Debt Management Matter? 41

we thus report the covariances between the nominal yield on corporate stock, the nominal yield on long-term government bonds, and the consumer price index. The corresponding co-efficients of correlation are reported in Table 4.2(*b*). Here we see that both long-term bonds and corporate stock have a negative correlation with the real asset, while they are positively correlated with each other (cf. Bodie 1982).

Using the covariances of Table 4.1, we can now apply formula (10) to calculate the policy effect. Assuming that the degree of relative risk aversion *c* is equal to 4, we have

$$\frac{\partial r_1^c}{\partial \alpha_2^s} = 0.00352 \quad \text{and} \quad \frac{\partial r_2^c}{\partial \alpha_2^s} = 0.01444. \tag{11}$$

This means that marginally increasing the stock of long-term bonds (α_2^s) would increase the equity yield (r_1^c), thereby 'crowding out' equity-financed investments. The figure 0.00352 might seem rather small, but compared with similar figures in the literature it is actually quite large. Our figure indicates that, if we increase the share of long-term bonds in the investor's portfolio by 1 percentage point (say, from 15 to 16 per cent, which corresponds roughly to the actual US portfolio share in 1987) in exchange for short-term debt, this would raise the quarterly yield on equity by 0.00352 of a percentage point. The increase in the yearly yield would then be four times as large, i.e. 0.0141 of a percentage point. For the same experiment within the framework of a quite different model and data set, Frankel (1985: 1057) obtains an increase in the yearly yield on corporate equity of only 0.0005 of a percentage point. Our figure is thus almost thirty times as large as that of Frankel. On the other hand, Roley (1982: 662) obtains a figure which for some periods is even larger than ours. Although these studies differ with respect to estimation period as well as model construction, they still indicate the wide range of possible policy responses.

Turning to the own-yield effect, the impact of debt management is more substantial. The policy derivative for long-term debt is thus 0.01444, or more than four times as large as the figure for equity. This means that a one percentage point increase in the portfolio share of long-term debt raises the expected yearly yield on long-term debt by 0.05776 of a percentage point. This

42   *Jonas Agell and Mats Persson*

figure is almost seven times as large as that reported by Frankel, who performs the same experiment.

4.2 A Vector Autoregression Approach

The figures in Tables 4.1 and 4.2 indicate the degree of covariation between corporate stock, long-term bonds, and short-term bonds (and some real asset represented by the consumer price index). One should however be careful not to mistake variability for uncertainty. For example, assume that a variable develops over time as the sum of two components: a completely predictable cycle (for example the business cycle) and a random disturbance from day to day (say the impact of the weather). Such a variable is shown in Fig. 4.1, where the predictable cycle is assumed to have a low frequency and a high amplitude, while the random disturbance is assumed to have a high frequency and a low amplitude.

Now, such a series would yield a rather large variance, owing to the high amplitude of the business cycle. But since this is

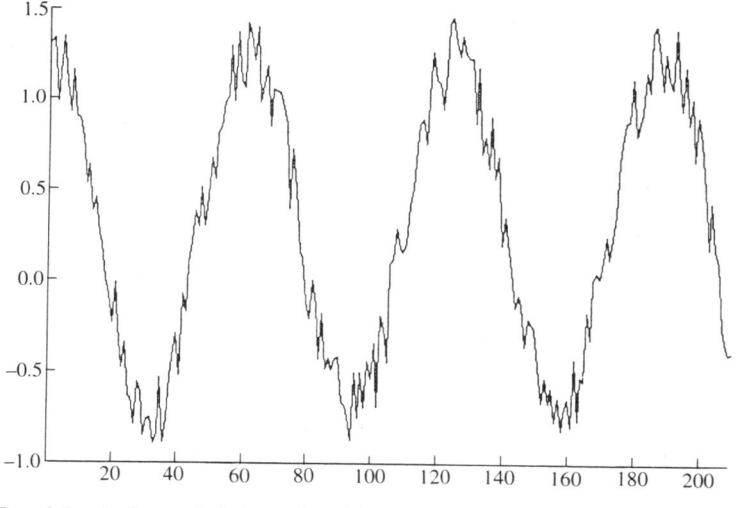

Fig. 4.1 A deterministic cycle with random disturbances

Does Debt Management Matter? 43

completely predictable, the large variance does not correspond to a high degree of uncertainty. The only source of uncertainty is the small random disturbance, and thus the variance of the series itself is an incorrect measure of the degree of uncertainty involved.

A way to cope with this is to estimate the coefficients of the ARMA process describing the time series of Fig. 4.1 (see Box and Jenkins 1970). Writing the time series as

$$X_t = k_0 + \sum_{i=1}^{\infty} a_i X_{t-i} + \varepsilon_t,$$

we have decomposed it into a predictable component ($k_0 + \sum a_i X_{t-i}$) and a random component (ε_t). More generally, if we have three different series X_t, Y_t, and Z_t, like the yields of our three assets above, we can run a vector autoregression according to

$$X_t = k_0 + \sum a_i X_{t-i} + \sum b_i Y_{t-i} + \sum c_i Z_{t-i} + \varepsilon_t \qquad (12)$$

$$Y_t = f_0 + \sum g_i X_{t-i} + \sum h_i Y_{t-i} + \sum m_i Z_{t-i} + e_t \qquad (13)$$

$$Z_t = n_0 + \sum p_i X_{t-i} + \sum q_i Y_{t-i} + \sum r_i Z_{t-i} + \eta_t. \qquad (14)$$

Having estimated the coefficients (k_0, a_i, b_i, c_i, f_0, g_i, h_i, m_i, n_0, p_i, q_i, r_i), one could say that all of the predictable variation has been removed, and that the true uncertainty that should be taken into account in the investor's portfolio decision is captured in the estimated covariance matrix of the residuals (ε_t, e_t, η_t).[4] This also means that the expected yields r^e can be more accurately computed than by using only the coupons of the three assets, since the system (12)–(14) provides us with optimal forecasts:

$$X_t^e = \hat{k}_0 + \sum \hat{a}_i X_{t-i} + \sum \hat{b}_i Y_{t-i} + \sum \hat{c}_i Z_{t-i}$$

$$Y_t^e = \hat{f}_0 + \sum \hat{g}_i X_{t-i} + \sum \hat{h}_i Y_{t-i} + \sum \hat{m}_i Z_{t-i}$$

$$Z_t^e = \hat{n}_0 + \sum \hat{p}_i X_{t-i} + \sum \hat{q}_i Y_{t-i} + \sum \hat{r}_i Z_{t-i}.$$

---

[4] This is not entirely true, since there is also some uncertainty regarding whether the estimated parameter values are equal to the actual parameter values. Since we can never know *a priori* whether there has recently been a major change in the stochastic process governing the development of the yields, it is not self-evident that the covariance of the residuals of the system (12)–(14), estimated from historical data, gives a better picture of the 'true' covariance matrix than the simple, unconditional matrix reported in Table 4.1. Cf. the discussion of the GARCH approach at the end of Sect. 4.3 below.

44   *Jonas Agell and Mats Persson*

Estimating the system $(12)-(14)$ on our data for real capital gains $(g - \dot{p})$ on shares, long-term bonds, and short-term bonds, and computing the covariances of the residuals gives us the covariance matrix displayed in Table 4.3.[5] We see that these 'conditional' variances are much smaller than the 'unconditional' variances reported in Table 4.1. This is a consequence of the fact that we have eliminated the predictable variation and preserved only the 'genuine' uncertainty in the time series. In particular, the uncertainty about future inflation (i.e. the variance of the real yield on short-term bonds) is virtually zero on a quarterly basis. However, the general features of Table 4.1 and 4.3 remain the same: the unconditional covariances of the former are positive, like the conditional covariances of the latter. And the coefficients of correlation are of the same order of magnitude in both tables, displaying the same pattern: the correlation between long-term bonds and short-term bonds is the highest, while the correlation between shares and short-term bonds in both tables takes an intermediate value.

TABLE 4.3 *Conditional covariances between the quarterly real yields of (1) corporate equity, (2) long-term government bonds, and (3) short-term government bonds: US data, 1960(1)–1988(2)*

		Corporate stock	Long-term bonds	Short-term bonds
(a)	*Covariance matrix*			
	Corporate stock	0.00315	0.00071	0.00008
	Long-term bonds		0.00315	0.00009
	Short-term bonds			0.00002
(b)	*Coefficients of correlation*			
	Corporate stock	1	0.22	0.31
	Long-term bonds		1	0.41
	Short-term bonds			1

Source: IMF, *International Financial Statistics*

[5] In our vector autoregression we have used a 4-period lag throughout. This lag structure, and the implied pattern of asset return dynamics, obviously does not fit easily with the hypothesis of efficient capital markets, which in its weak form suggests that past price information is unrelated to future prices.

4.3 Changes in the Information Set

Eliminating predictable variation from the data makes it possible for us to concentrate on the uncertainty underlying the investor's portfolio choice. We have not yet, however, clearly defined his *information set*. To be able to make the estimates underlying the residual covariances reported in Table 4.3, the agent would have to use quarterly data from the entire sample period, i.e. from 1960(1) to 1988(2). At some intermediate date, say 1975(3), not all this information is available to him: the best he can do then is to make estimates using the data from 1960(1) to 1975(2). This means that the parameter estimates of equations (12)–(14) made at 1975(3) will be different from those underlying Table 4.3, and thus the agent's perception of the covariance matrix at 1975(3) will also be different. The agent's perception of the covariance matrix will hence vary over time, as more and more data points become available to him. We have taken this into account in the same fashion as Friedman (1985*b*, 1986) has done.

Our data series begins in 1960(1), and we assume that the agent's data series does the same. At 1970(1) enough observations are available to permit reasonably reliable vector autoregressions of the system (12)–(14). For each date following 1970(1) we have re-estimated the system, recomputed the residuals, and recalculated the corresponding covariance matrix. Thus there will be a new covariance matrix for each quarter following 1970(1). One would expect that these matrices will not change much from one quarter to another, since the parameter estimates of the system (12)–(14) will not change very much when only one more observation is added. Still, the changes can sometimes be quite substantial, as is shown in Fig. 4.2. In particular, the sharp jump in the variance of long-term bonds in 1982 is surprising; it depends on the sharp decline in the interest rate between June and October that year, when the long-term interest rate fell by approximately three percentage points.[6] We also see that, while there seems to be no clear trend in the variance of corporate stock, the variance of long-term debt has risen almost

---

[6] The decline in the short-term interest rate was even more dramatic. As a result of the easing up of the monetary policies of the Federal Reserve in the summer of 1982, the Treasury bill rate fell by 5.5 percentage points between June and August.

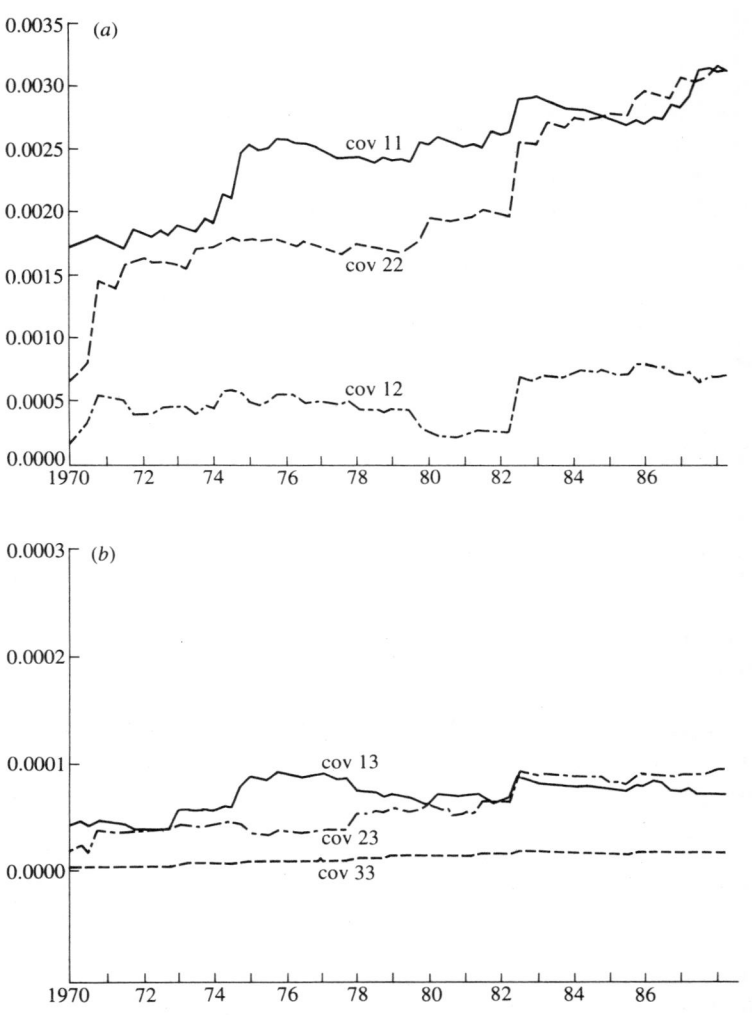

FIG. 4.2 Conditional quarterly covariances between the real yields on (1) corporate stock, (2) long-term bonds, and (3) short-term bonds, 1970(1)–1988(2)

Does Debt Management Matter? 47

monotonically since 1979. The covariances and the variance σ_{33} (real yield on short-term bonds, i.e. inflation risk), however, seem to be fairly stationary. The numerically low levels of the covariances including short-term bonds are also striking: the maximum values of σ_{13}, σ_{23}, and σ_{33} are substantially lower than the minimum value of σ_{12}. This indicates that the vector autoregression model (12)–(14) is rather efficient in explaining inflation, and it would not be unreasonable to treat short-term bonds as a safe asset also in real terms. Finally, the coefficients of correlation are shown in Fig. 4.3. Their development is rather stationary, and they are of approximately the same order of magnitude as the unconditional ones reported in Table 4.1(b) above.

The sequence of covariance matrices obtained by the adaptive vector autoregression method has been plugged into the expression for the policy derivatives (10). This results in a sequence of policy derivatives $\partial r_1^c/\partial \alpha_2^s$ and $\partial r_2^c/\partial \alpha_2^s$, as shown in Fig. 4.4. We see first that both derivatives are positive, thus indicating that an increase in long-term financing of government debt will increase the cost of capital for the private sector, thereby reducing

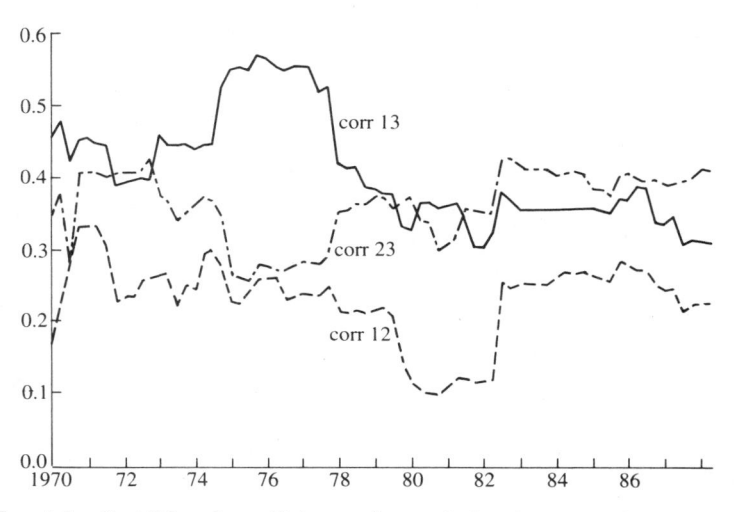

FIG. 4.3 Conditional coefficients of correlation between the quarterly yields on (1) corporate stock, (2) long-term bonds, and (3) short-term bonds, 1970(1)–1988(2)

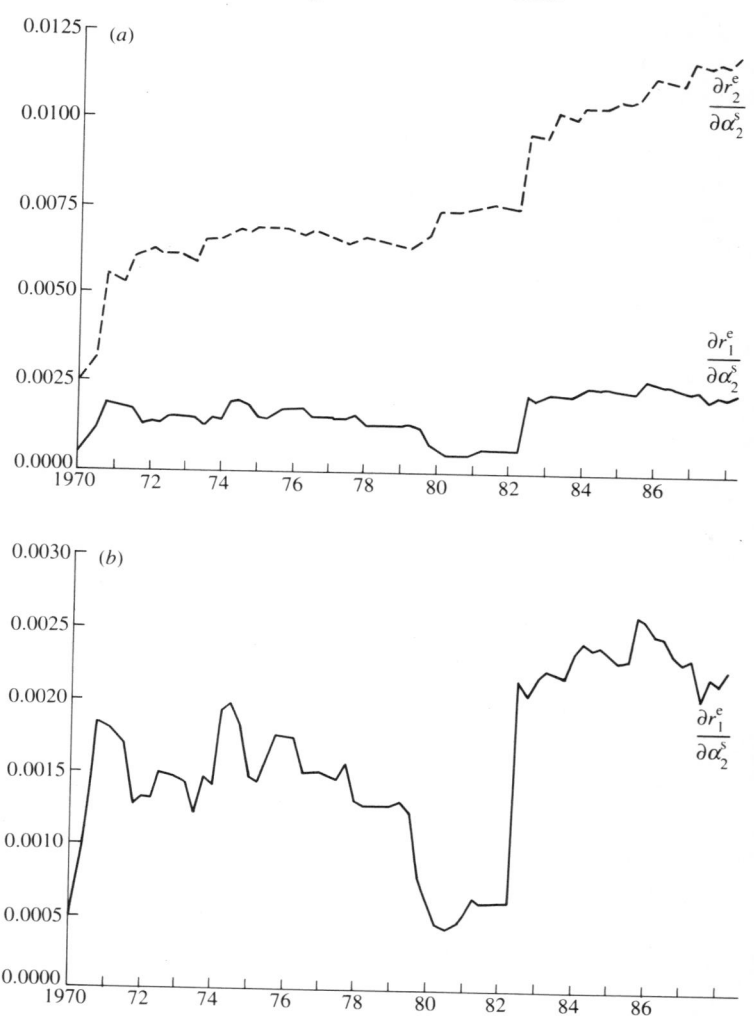

FIG. 4.4 The effects on asset yields of increasing the supply of long-term bonds: quarterly data, 1970(1)–1988(2): (a) both policy derivatives; (b) the derivative $\partial r_1^e / \partial \alpha_2^s$ on a magnified scale

Does Debt Management Matter? 49

investment in industry. Equivalently, a change to more short-term financing will tend to stimulate industrial investment. It is worth noting that the signs of the two derivatives are the same over the entire period, which could be regarded as a sign of the robustness of the result. If $\partial r_1^c / \partial \alpha_2^s$ had changed in sign frequently over the period, one would be less inclined to believe firmly in the qualitative result that more short-term financing of government debt stimulates investment.

Second, we see that, while there is no obvious trend in the development of $\partial r_1^c / \partial \alpha_2^s$, there is a marked upward trend in $\partial r_2^c / \partial \alpha_2^s$. In fact, however, the path of $\partial r_1^c / \partial \alpha_2^s$ shows violent variations over time, from a minimum of 0.00042 in 1980(3) to a maximum of 0.00260 in 1985(3). This is an increase of 500 per cent. To emphasize this, we have plotted $\partial r_1^c / \partial \alpha_2^s$ separately, using another scale, in Fig. 4.4(*b*). One could therefore say that, although the qualitative result (i.e. the positive sign of the derivative) is robust, quantitative results should be regarded with caution.

Translating the quarterly yield effects into yearly figures, we see that an increase in long-term debt by one percentage point (from 15 to 16 per cent of total wealth) increases the yearly yield of corporate stock by a minimum of 0.00168 and a maximum of 0.0104 of a percentage point. Of these, the latter is perhaps the more interesting, since it refers to a later date in the estimation period. Our figures are considerably lower than the corresponding ones obtained using unconditional covariances reported in (11) above. The maximum figure (obtained for 1985(4)) is about 30 per cent smaller than the unconditional derivative, while the minimum figure (1980(3)) is eight times smaller than the unconditional one. Also, the minimum figure is about three times as large as that of Frankel (1985: 1057) while the maximum is about 20 times as large, thereby emphasizing the violent changes in policy responses over time.

The own-yield effect $\partial r_2^c / \partial \alpha_2^s$ increases in a trend-like manner, from a minimum of 0.0025 in 1970(1) to a maximum of 0.0119 in 1988(2). The last figure is about 20 per cent smaller than the unconditional derivative reported in (11). In yearly terms, increasing the share of long-term bonds by one percentage point will, using the conditional covariances for 1988(2), raise the expected bond yield by a mere 0.0476 of a percentage point—a

50 *Jonas Agell and Mats Persson*

figure that is almost six times larger than that reported by Frankel, but still substantially smaller than some of the simulation results reported by Roley (1982) and Friedman (Part II below).[7]

What policy conclusions can we draw from the preceding analysis? In qualitative terms, the results conform with standard presumptions. Lengthening the maturity composition of government debt increases the expected yields on corporate equity and long-term bonds relative to the expected return on short-term debt. For a given expected return on short-term debt, this means that an increase in the supply of long-term government bonds (in exchange for Treasury bills) increases the costs of corporate financing and 'crowds out' corporate investment. A full discussion of whether the calculated relative return adjustments are large enough to matter *economically* would of course require bringing in additional macroeconomic structure concerning aggregate supply and demand relationships. A more casual inspection, however, suggests that the effects are small. Using the conditional covariances computed for 1988(2), we find that increasing the share of long-term bonds in investors' portfolios by (an unprecedentedly large) 10 percentage points will raise the expected yearly yields on corporate equity and long-term bonds by about 0.09 and 0.48 of a percentage point, respectively.

Also, in assuming that the government can determine relative asset supplies at its own discretion, we have implicitly assumed that the capital structure decisions of firms are exogenously given. More realistically, the ultimate effects of debt management operations depend on the extent to which firms respond by adjusting their liability mix. As lengthening the maturity composition of government debt drives up the relative yields on corporate equity and long-term bonds, we would expect firms to rely less on financing by long-term bonds and corporate equity and more on financing by short-term bonds. As these private supply adjustments at least partly neutralize the debt management operations of the government, the net impact of

---

[7] Also, using a vector autoregression model similar to ours, Friedman (1985*b*) concludes that debt management may affect relative asset yields in an economically significant way. Friedman's results, however, are not directly comparable with ours, as they, by pertaining to *new* debt issues rather than pure open-market operations, also reflect wealth effects induced by an increase in net private wealth. (Cf. the discussion in Sect. 2.1 above.)

Does Debt Management Matter? 51

debt management on asset yields will typically be even less significant than the yield effects reported above.

As repeatedly stressed, systematic use of debt management also requires stable and predictable yield responses in asset markets. As indicated by our results, this requirement creates additional difficulties for the authorities managing the government debt. In particular, we found that the policy derivative for the return on corporate equity (i.e. the target variable most commonly associated with debt management in the literature) exhibited sharp fluctuations over time, thus underlining the difficulty of using debt management for fine-tuning purposes.

Implicit in the approach employed so far above is a particular view of the world. It is assumed that the stochastic processes governing asset yields are stationary, i.e. that the elements of the underlying 'true' covariance matrix are constant. The agent's perception of the covariance matrix changes as more data points become available, but the objective probability distribution actually generating the data is assumed to be time-invariant.

An alternative view would be to allow for non-stationarity in the underlying stochastic processes. This could be motivated by, for example, changing monetary regimes—as governments, central bankers, and doctrines of stabilization policy come and go, so does the nature of surprises confronting investors in asset markets. It could also be motivated by shifts in the underlying asset technology; such shifts (e.g. the development of new financial instruments) may occur at random intervals and thereby affect the return pattern until the next shift occurs. Of course, the basic stochastic process could still be stationary, and could be defined as the joint outcome of random changes in returns, in monetary regimes, and in technology; describing the asset yields by a stationary autoregressive system like (12)–(14) would then still be appropriate. Nevertheless, it could sometimes be suitable to estimate the asset return structure using methods that explicitly allow for non-stationarity.

The simplest way to do this is to use the 'depreciation method' developed by Friedman and Kuttner (1988). Since older observations are more likely to have been generated by a different process than more recent ones, old data points are given less weight in the estimation of the system (12)–(14). Another approach is to use the so-called Generalized Auto-Regressive

52   *Jonas Agell and Mats Persson*

Conditional Heteroskedasticity (GARCH) estimation technique, developed by Engle (1982) and applied to asset markets by e.g. Bollerslev, Engle and Wooldridge (1988). With this approach, the elements of the 'true' covariance matrix are themselves regarded as being generated by some autoregressive process, and the technique allows for estimating the parameters of that process.

There is no a priori reason to prefer one of these approaches to the others. The proper choice between them has to be based on practical considerations and on extensive empirical experience (cf. Friedman and Kuttner 1988). We have here chosen to work with the simplest assumption, i.e. that of a stationary underlying stochastic process, but we want to indicate the vast opportunities for further empirical research to shed light on this issue.

4.4   The Time Aggregation Problem

The policy conclusion of the previous section is straightforward, at least qualitatively. It does however rely on the data material used. We, as well as many other students of debt management referred to above,[8] have used quarterly data. The question is whether the choice of the data interval affects the conclusion.

To illustrate the problem, we have constructed a hypothetical example. Assume that two time series X and Y move as depicted in Fig. 4.5. Computing their covariance yields cov (X, Y) = −0.689. Let us now however use a different periodization, taking only every second observation into account. The two time series thus obtained look rather different, and their covariance is 0.124. Taking instead every third observation into account yields cov (X, Y) = −0.434. In Fig. 4.6 we show the values of the covariance between X and Y for different choices of the data interval. We see that not even the sign of the covariance is robust to the choice of data interval, let alone the magnitude. Particularly striking are the very large values obtained for a period length of 11 and 13, emphasizing the high sensitivity of the estimates of a covariance matrix Ω.

---

[8] e.g. Friedman (1985*b*, 1986) and Roley (1982). Frankel (1985) used yearly data for his study.

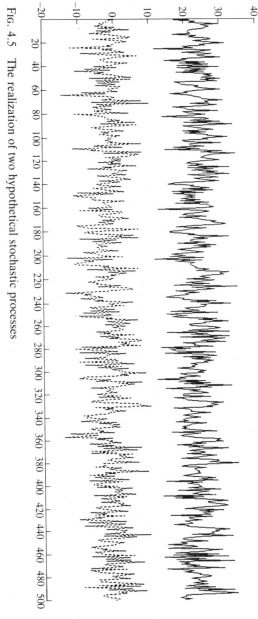

Fig. 4.5 The realization of two hypothetical stochastic processes

54   *Jonas Agell and Mats Persson*

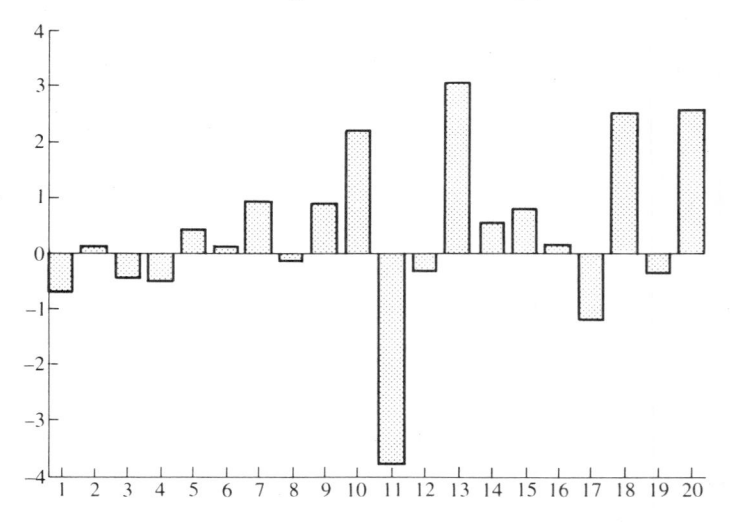

FIG. 4.6   The covariances of the two time series in Fig. 4.5 for different data intervals

The econometric problems involved in choosing the data interval have been a neglected issue in most of the literature. Some studies have been made in the field of marketing research,[9] showing how the quantitative conclusions regarding the impact of various marketing actions depend on whether monthly, quarterly, or yearly data are used.

What then is the 'proper' data interval? That of course depends on the nature of the agent's decision problem. In our single-period portfolio balance model, the data interval should be equal to the investor's holding period, which is the time interval between successive portfolio reallocations. But the holding period varies over assets and agents. If we had included real estate among our assets, the planning horizon for, say, owner-occupied homes should perhaps be measured in years if not decades, owing to the considerable transaction costs involved, while real estate

---

[9] See the references given in Berndt (1991). In the financial economics literature, Grossman *et al.* (1985) have discussed some of the problems involved when estimating continuous-time asset pricing models using data that are time-averaged. For a discussion of the problem from the point of view of statistical theory, see Bergstrom (1984).

Does Debt Management Matter?   55

bought by professional investors might be associated with a shorter holding period. For bonds and shares traded by tax-exempt institutional investors, the holding period could be as short as a few minutes. For bonds and shares bought by households, the transaction costs and the tax rules involved may imply a holding period of, say, a few months or perhaps a year or two.[10]

A complete model would determine the optimal holding period together with the optimal portfolio demands. However, when we try to represent a multi-period world with transaction costs and heterogeneous investors by a simple model of the type used here, it is impossible to say a priori what is the correct periodization of the data series. What one can do is to replicate the analysis for different data intervals and see whether the policy conclusions are robust with respect to the choice of data interval.

In Fig. 4.7 we therefore report the covariances, and in Fig. 4.8 the coefficients of correlation, using monthly data, that result from the vector autoregression model of equations (12)–(14). When comparing those covariances with their quarterly counterparts in Fig. 4.2 above, we see that the general shapes of the curves are quite similar.[11] The important difference is in the order of magnitude: the monthly variances are about three times as large as the quarterly ones.[12] The coefficients of correlation reported in Fig. 4.8 differ somewhat from their quarterly counterparts of Figure 4.3. Apart from the order-of-magnitude problem, we see that ρ_{12} is less than both ρ_{13} and ρ_{23} over the entire sample period when we use quarterly data, but that ρ_{12} is larger than ρ_{23} at the beginning of the period when we use monthly data.

Now, it is hard to judge from a visual inspection whether or

---

[10] Cf. Fischer (1983) for a discussion of how the dynamics and uncertainty of asset returns depend on the holding period of investors.

[11] To make the one-month interest rates comparable with the quarterly interest rates, we have multiplied the former by a factor of 3. Further, when estimating the model (12)–(14) in Section 4.2 above using quarterly data, we assumed a 4-period lag; when using monthly data, we have consequently assumed a 12-period lag.

[12] This might depend on the construction of the basic data. They refer to period averages, and—depending on the stochastic properties of the time series—the averaging procedure might have the effect that the variance becomes smaller for long periods (e.g. a quarter) than for short periods (e.g. a month.).

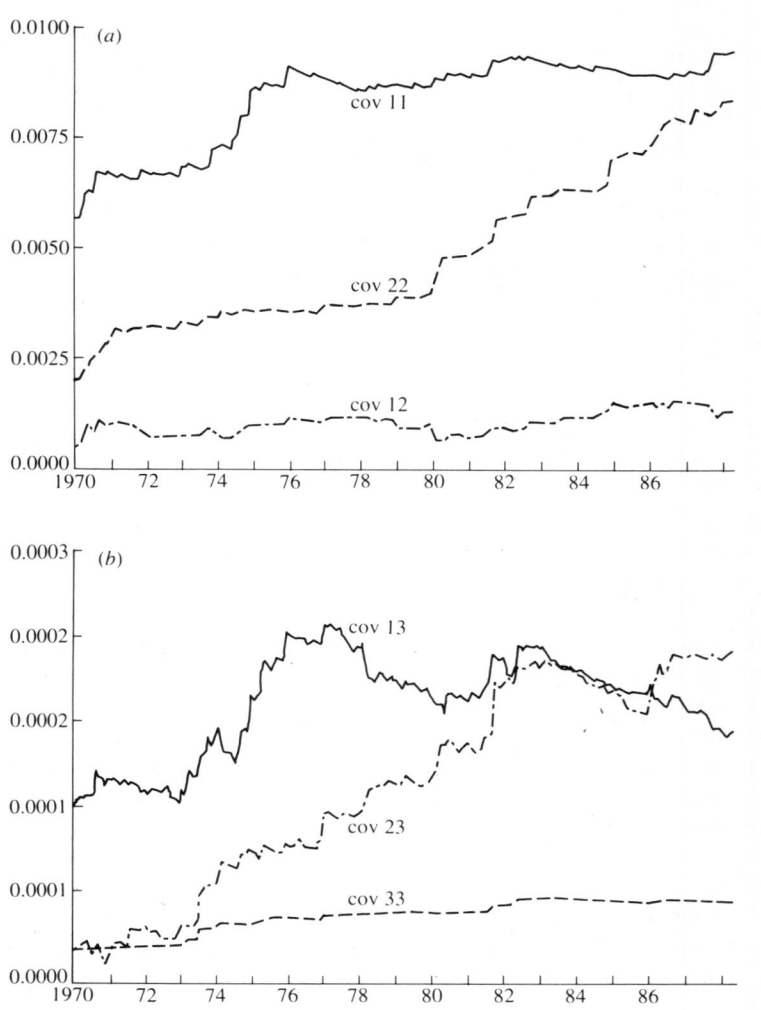

Fig. 4.7 Conditional monthly covariances between the real yields on (1) corporate stock, (2) long-term bonds, and (3) short-term bonds, 1970(1)–1988(2)

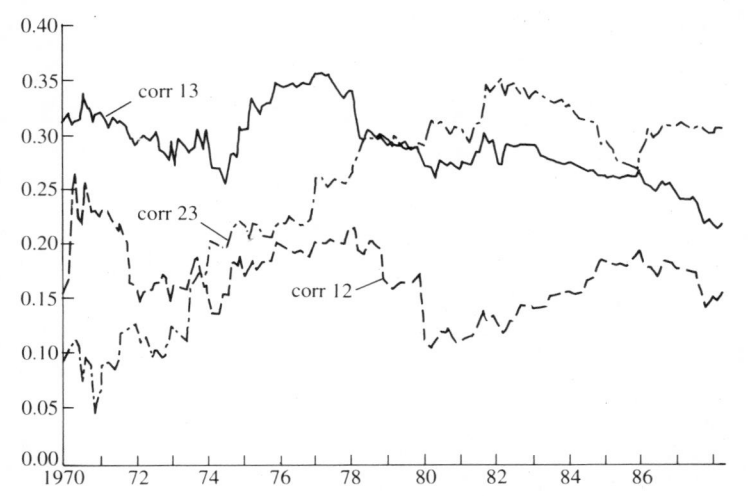

FIG. 4.8 Conditional coefficients of correlation between the real yields on (1) corporate stock, (2) long-term bonds, and (3) short-term bonds: monthly data, 1970(1)–1988(2)

not these deviations in pattern are important. We have therefore recomputed the 'policy derivatives' of (10) above using monthly data. In Fig. 4.9(a) we have plotted $\partial r_1^c/\partial \alpha_2^s$, i.e. the impact on the equilibrium yield on corporate stock of an increase in the supply of long-term bonds (the solid curve) using monthly data. As a comparison, we have also plotted the same derivative using quarterly data (the dotted curve: this is the same curve as that in Fig. 4.4 above). We see that the qualitative properties are the same, indicating crowding-out. The peak in 1985–6, the trough in 1980–1, and the varying pattern in the 1970s are also the same. The orders of magnitude are however different; the crowding-out effect on the equity yield seems much stronger if we use monthly data. As is evident from Fig. 4.9(b), similar considerations apply to the derivative $\partial r_2^c/\partial \alpha_2^s$.

In sum, although the qualitative conclusions seem robust with respect to the choice of data interval, the quantitative conclusions seem much more sensitive. Any empirical conclusion concerning the potency of debt management may thus depend crucially on the time length between the data observations available to the researcher.

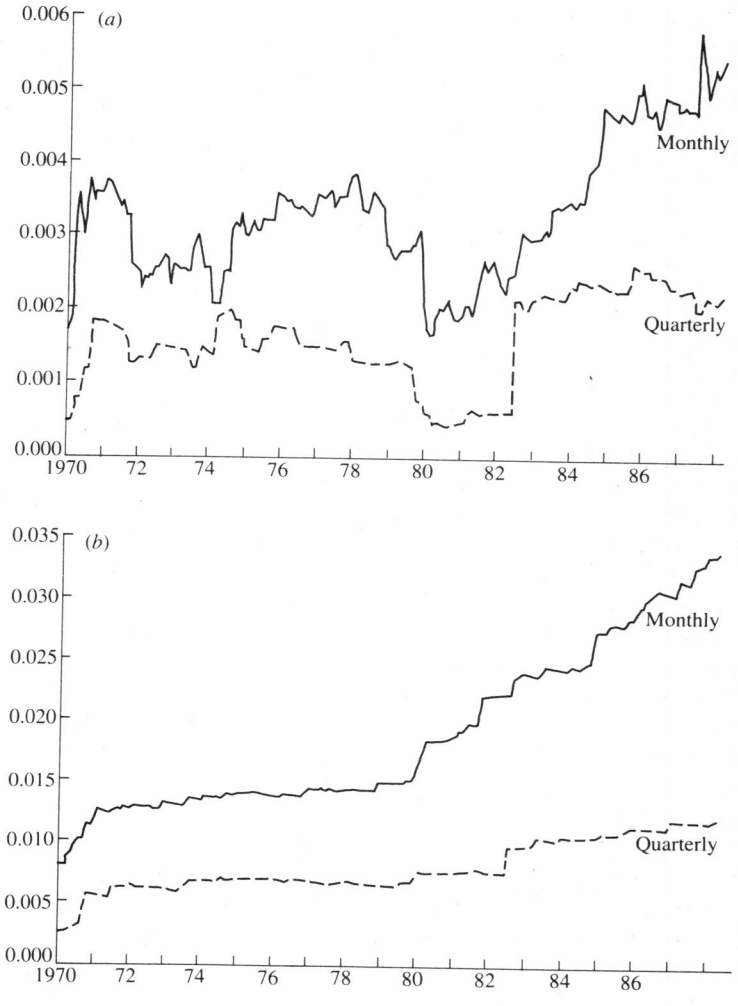

FIG. 4.9 (a) The derivative $\partial r_1^c/\partial\alpha_2^s$ for monthly and quarterly data, 1970–1988
(b) The derivative $\partial r_2^c/\partial\alpha_2^s$ for monthly and quarterly data, 1970–1988

5

An Alternative Approach to the Covariance Matrix

In the previous analysis we have considered the covariance structure as something objectively measurable that can be constructed out of historical data. Either it is simply the unconditional covariance matrix of Table 4.1 above, or it is the conditional covariance matrix of Table 4.3. On a more sophisticated level, it is the 'adaptive' structure reported in Figure 4.2, which allows risk perceptions to vary over time.

In reality, however, the problem is much more difficult. As the risk perceptions of investors are inherently subjective, it is far from self-evident that the 'true' elements of the covariance matrix Ω coincide with the ones we have estimated from historical data. In the real world we would thus expect investors to use whatever information they have available when forming their risk-return beliefs. In addition to the backward-looking information implicit in historical return data, their information sets may include 'news' in the form of recent announcements by the government of drastic changes in monetary policy, tax rules, or debt management policies, as well as 'noise' in the form of market rumours unrelated to changes in economic fundamentals.

This problem is also inherent in the GARCH approach of Engle (1982) and Bollerslev et al. (1988) referred to above. Such an approach would employ more sophisticated estimation techniques than the ones used in the previous chapter in the sense that it also includes estimation of how the elements of the actual (as opposed to the perceived) covariance matrix change over time. However, it nevertheless uses a generalized vector auto-regression methodology, thereby confining itself to the use of historical data for computing the covariances.

There are in principle several escapes from the adaptive expectations straightjacket implicitly underlying the vector autoregression procedure used in the previous chapter. A first

60   *Jonas Agell and Mats Persson*

possible solution is the rational expectations method developed by Frankel (1985). In every period t, Frankel assumes that the expected returns entering the asset demands of investors equal the realized *ex post* returns plus an error term uncorrelated with any information available to investors at time $t - 1$. Thus, the expected returns can vary freely over time and are not restricted to any particular backward-looking expectations mechanism. Finally, by imposing the constraint that the subjective covariance matrix Ω equals the covariance matrix of the residuals associated with estimating an equilibrium asset market model of the form (9), Frankel obtains estimates of the relevant asset demand parameters.

A second method is to use the information contained in opinion surveys. This approach is represented by Friedman (1986), who infers investors' risk perceptions from expectational survey data concerning inflation, stock prices, and long-term interest rates. Such a survey data methodology is of course subject to the standard criticism that we do not know a priori whether the people interviewed really are identical, or are even remotely similar, to the representative individual(s) in the market. Still, the survey approach is warranted; the elusiveness of the very concept of expectations calls for considerable eclecticism in empirical work. We simply have to try all possible approaches in order to get some view of the robustness of the empirical results.

A third procedure is suggested in this chapter. For the diagonal elements σ_{ii} in the covariance matrix, we will use the subjective variances of stock and bond yields that can be inferred from options data and standard models of option pricing.[1]

According to the Black and Scholes (1973) options pricing formula, the price of an option is a rather complicated function of today's price of the underlying asset, the variance of that price, the time till the option can be exercised, the strike price, and the risk-free interest rate. That formula has been used mainly to calculate theoretical option values, but of course it could also be used the other way around: with knowledge of the market price actually paid for the option, and of all the other variables except the variance, one can compute the variance implicit in the observed market price.

---

[1] This possibility was first suggested to us by Jeffrey Frankel.

Does Debt Management Matter? 61

For the stock market, we have used the Standard & Poor's 500 Index, of which an option is traded at the Chicago Mercantile Exchange. For the long-term government bonds we have used the option traded at the Chicago Board of Trade, for which the underlying asset is a futures contract on a 15–20-year Treasury bond. There are also options for which the underlying asset is the Treasury bond itself; but these markets are rather thin, and it is not self-evident that the prices quoted are the 'correct' ones. The options on futures contracts, however, are very actively traded, and it seems fairly reasonable to use a theoretical option pricing formula when studying this market. Such an option is most appropriately evaluated using the Black (1976) formula, which is the one we have employed in this context. Since the underlying bond is of such a long duration, and since we deal only with options with a very short time duration (i.e. a quarter), the problems of compound interest, and the approaching date of redemption can be disregarded.

We have used quarterly data for the period 1985(4)–1988(2), and we have used the price quotations reported in the *Wall Street Journal* at the end of each quarter. The options studied have been those with a strike price closest to the current market price of the underlying asset, and with three months left to the strike date. As a measure of the risk-free interest rate, we have used the current yield on three-month Treasury bills. The implicit variances thus obtained are reported in Fig. 5.1(*a*) for the stock market and Fig. 5.1(*b*) for the bond market.

These series of variances have several interesting features, in particular if we compare them to the series based on historical yield data as reported in Fig. 4.2 above. For comparison, the series for the period 1985(4)–1988(2) are also depicted in Fig. 5.1 in the form of broken curves. We see that the variances computed from options data (the implicit variances) are considerably lower than the variances based on vector autoregressions on historical data (the VAR variances). This is what one would expect. If historical yield data are available to the agents in the options market—which they presumably are—then the information set underlying the implicit variances is larger than that underlying the VAR variances. And thus the latter should be larger than the former.

An exception is provided by the stock market crash in the

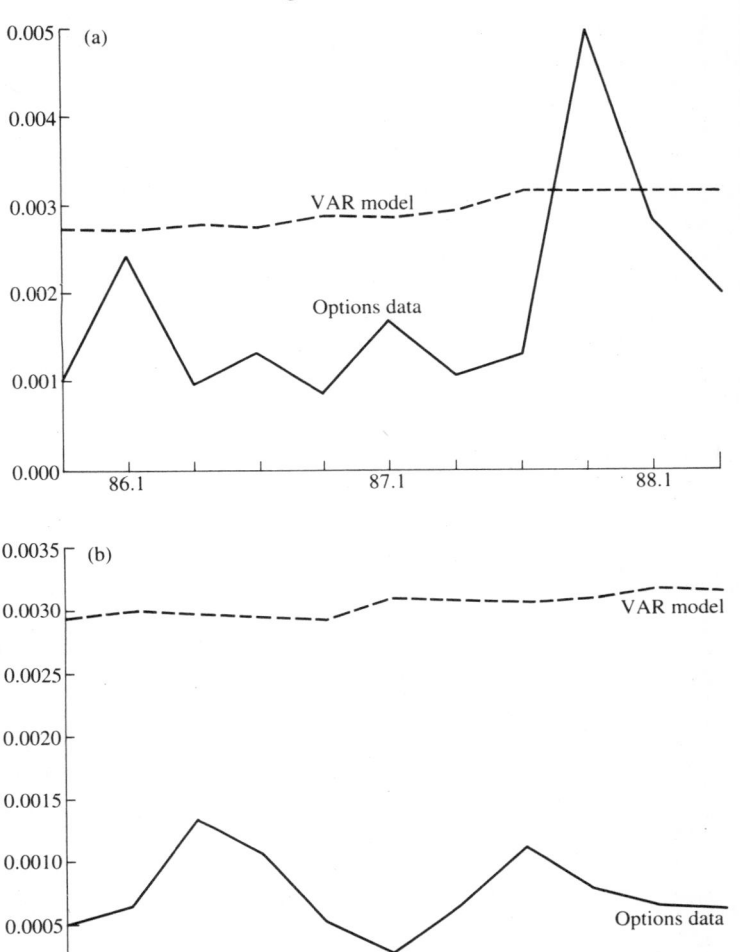

FIG. 5.1 A comparison of the quarterly variances of nominal asset yields using options data versus autoregression procedures, 1985(4)–1988(2): (*a*) variance of corporate equity; (*b*) variance of long-term bonds

Does Debt Management Matter?

fourth quarter of 1987. Here we have an event that was absent in the historical data, and because of the confusion and uncertainty during these hectic days the implicit variance increases drastically while the VAR variance reacts hardly at all. Soon, however, the implicit variance resumes its old level, while the VAR variance slowly increases far into 1988. Finally, we see that, interestingly enough, the variance of bond yields is totally unaffected by the dramatic events in the stock market, both when we use the implicit approach and when we use the VAR approach.

To translate the variances of Fig. 5.1 into numerical values of the policy coefficients $\partial r_i^e/\partial \alpha_j^s$, we recall from Chapter 3 above that, in the case of three assets, where one is riskless, the derivative can be written

$$\frac{\partial r_i^e}{\partial \alpha_j^s} = c\sigma_{ij}, \qquad i,j = 1,2. \tag{10'}$$

From the options data, we have values of σ_{11} and σ_{22}, and since

$$\sigma_{12} = \rho_{12}\sqrt{\sigma_{11}\sigma_{22}}, \tag{15}$$

we can compute the covariance σ_{12} if we know the correlation coefficient ρ_{12}. Here we have to rely on historical data. In Fig. 4.3 above we reported the conditional quarterly correlation coefficients based on the vector autoregression technique. For the period 1985(4)–1988(2), these numbers have been plugged into formula (15), together with the implicit variances σ_{11} and σ_{22} obtained from options data. The resulting values of the policy parameter $\partial r_1^e/\partial \alpha_1^s$ for that period are shown in Fig. 5.2(a).

For the policy derivative $\partial r_2^e/\partial \alpha_2^s$, we do not need any correlation coefficient ρ_{12}. Thus we do not have to resort to any figures obtained by VAR techniques, but can compute $\partial r_2^e/\partial \alpha_2^s$ directly from the variances implicit in bond option prices. This policy derivative is shown in Fig. 5.2(b). For comparison, the policy derivatives obtained from historical data are also shown in Fig. 5.2 by the broken curves.[2]

[2] These are in principle identical to those shown in Fig. 4.4 above, for the period 1985(4)–1988(2). In practice, however, there is a slight and hardly observable difference: the numbers in Fig. 4.4 were computed on the basis of the assumption that all three assets (including Treasury bills) are risky, while when computing the numbers in Fig. 5.2 we assumed that two assets are risky while Treasury bills are riskless. We have thus here disregarded the inflation risk, which in practice is negligible since the VAR technique reduces the uncertainty about inflation to almost zero (cf. Table 4.3 above).

64 **Jonas Agell and Mats Persson**

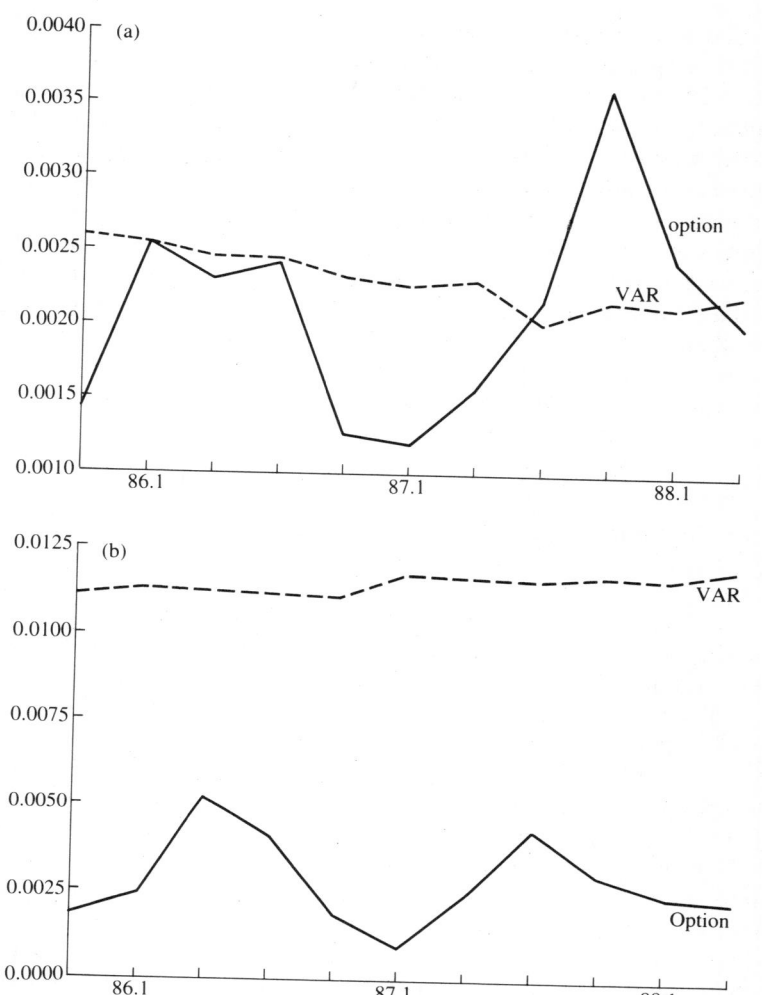

FIG. 5.2 (a) The derivative $\partial r_1^e/\partial \alpha_2^s$ for options and VAR approaches: quarterly data, 1985(4)–1988(2)

(b) The derivative $\partial r_2^e/\partial \alpha_2^s$ for options and VAR approaches: quarterly data, 1985(4)–1988(2)

Does Debt Management Matter? 65

Two features should be noted. First, we see that the policy derivative based on options data is much more volatile than the derivative based on historical data. This is what one would expect. The very concept of an autoregressive process implies a large degree of inertia in the time series, while the options data approach allows the variance to respond immediately to new information in the form of new policy announcements, more or less substantiated market rumours, etc. Second, except for the stock market crash in the fourth quarter of 1987, the derivatives based on options data are much lower than the ones based on historical data.

6
How Returns Adjust:
The Effects of Endogenous Prices

In the preceding chapter we discussed the effects of government debt management using a portfolio balance model widely referred to in the literature. Like all economic models, this 'work-horse' model can be characterized as a compromise between analytical tractability and economic plausibility. Some of the well-known simplifying assumptions include the suppression of intertemporal dependencies and taking the supply of financial assets as exogenously given. This chapter focuses on another, perhaps less obvious, analytical short cut. Most of the literature examines the effects of changes in the composition of government debt on expected asset returns while taking current asset prices as exogenously given. As noted by Friedman and Kuttner (1988: 26), this 'embodies the contradiction of implicitly taking as given the prices of the assets whose expected returns the model is supposed to determine, even though for most assets it is primarily variation in price that delivers variation in expected return'.

We will, in the following, incorporate endogenous adjustments of asset prices into the basic model. This entails introducing three distinct adjustment mechanisms, with potentially important implications for the effects of government debt management: (i) allowing for valuation changes implies introducing wealth effects; with endogenous prices, initial portfolio wealth W becomes an endogenous variable; (ii) the elements of the covariance matrix Ω now become endogenous; (iii) with given asset prices, any distinction between asset supplies defined in quantitative terms and in value terms is immaterial while with endogenous prices this distinction may be important. The third point says that, if we perform debt management in the sense of changing the supply of some asset in quantity terms, then the supplies of other assets may change in value terms—even if we have not changed the supplies of those assets in quantity terms.

Does Debt Management Matter? 67

This chapter provides qualitative and quantitative assessments of the effects of debt management when allowing for endogenous prices. Sections 6.1 and 6.2 describe the model and derive the relevant comparative-static results. Section 6.3 turns to the empirical evidence and compares the results of the 'work-horse' model used in Chapter 4 with those implied by the model incorporating valuation changes.

6.1 The Basic Model with Endogenous Asset Prices

The mechanism behind the change in some expected yield r_i^e is of course that the price of asset i changes. In our simple atemporal model, there is only one time period. We denote the price of the asset at the beginning of the period by P_i and at the end of the period by $\tilde{P}_i$, and let the tilde indicate that $\tilde{P}_i$ is uncertain. Disregarding the coupon (we could for example treat all assets as discount bonds—see e.g. Roley 1979), the yield r_i is defined by

$$r_i \equiv \frac{\tilde{P}_i - P_i}{P_i}$$

and the expected yield r_i^e is the corresponding mathematical expectation with respect to the probability distribution of $\tilde{P}_i$.

A change in the expected yield r_i^e, then, means that 'today's' price P_i changes, or that 'tomorrow's' expected price $\tilde{P}_i^e$ changes, or both. Without specifying the institutional setup of the model further, we cannot say which one applies; we can only say that the prices P_i and $\tilde{P}_i$ have to change in such a fashion that equation (7) is satisfied. In the case of short-term discount bonds, where the time to maturity is equal to our period of analysis, the future price $\tilde{P}_i$ is legally fixed in nominal terms,[1] and all changes in r_i^e have to be channelled via changes in today's price P_i.

However, in the case of long-term assets, with a time to maturity going beyond our period of analysis, it is natural to regard both P_i and $\tilde{P}_i$ as endogenously determined market prices. The exact relationship between the two prices is then ambiguous. A policy $d\alpha_j^s$, which, via the formula (10), leads to an increase in r_i^e, could have two effects on the price P_i. Either the increase in r_i^e is

---

[1] For the time being we disregard the problem of nominal versus real yields.

68   *Jonas Agell and Mats Persson*

accomplished by an increase in today's price P_i, together with a sufficiently large increase in tomorrow's price $\tilde{P}_i$ so as to make the ratio $(\tilde{P}_i - P_i)/P_i$ grow, or it is accomplished by a decrease in P_i, together with some (perhaps minor) change in $\tilde{P}_i$. The question of which alternative to apply can be settled only by the analysis of an explicitly intertemporal model. In practice, however, it is always implicitly assumed that r_i^e and P_i move in opposite directions; that is, an increase in the yield is achieved by a fall in today's price of the asset.[2]

Let us now start with the demand system (6) in value terms, with all asset prices explicit. Setting demand equal to supply, also in value terms, we have the equilibrium system

$$
\begin{bmatrix} P_1 q_1^s \\ P_2 q_2^s \\ \vdots \\ P_n q_n^s \end{bmatrix} = \left(\frac{1}{c} B \begin{bmatrix} (\tilde{P}_1^c - P_1)/P_1 \\ (\tilde{P}_2^c - P_2)/P_2 \\ \vdots \\ (\tilde{P}_n - P_n)/P_n \end{bmatrix} + \Pi \right) (P_1 \bar{q}_1 + P_2 \bar{q}_2 + \ldots + P_n \bar{q}_n) \quad (16)
$$

where $\tilde{P}_i^c \equiv E[\tilde{P}_i]$, and where q_i^s is the supply of asset i and $\bar{q}_i$ is the corresponding initial endowment. The matrix B and the vector Π are defined as before in terms of the covariance matrix Ω. The latter is given by

$$
\Omega \equiv [\text{cov}(r_i, r_j)] \equiv \{\text{cov}[(\tilde{P}_i - P_i)/P_i, (\tilde{P}_j - P_j)/P_j]\}
$$
$$
\equiv \left[\frac{1}{P_i P_j} \text{cov}(\tilde{P}_i, \tilde{P}_j) \right] \equiv \left[\frac{1}{P_i P_j} \tilde{\Omega} \right], \quad (17)
$$

where $\tilde{\Omega}$ is the covariance matrix of the end prices $\tilde{P}_i$. More compactly, (17) can be written as

$$
\Omega \equiv \Gamma \tilde{\Omega} \Gamma,
$$

where Γ is the $n \times n$ diagonal matrix with elements $1/P_i$ in the diagonal and zeros elsewhere.

Several things should be noted here. First, the left-hand side of (16) is the supply vector. Since it is written in value terms, it

---

[2] This question is crucial if we want to use debt management for the purpose of stabilization policy. If a policy action $d\alpha_j$ results in an increase in the equilibrium yield of common stock ($dr_i^e > 0$), we say that the policy leads to a reduction of industry investment demand. But if $dr_i^e > 0$ goes hand in hand with an increase in today's share prices ($dP_i > 0$), it means that Tobin's q has actually increased, which should stimulate investment demand.

Does Debt Management Matter? 69

includes the market prices P_i and the physical supplies (i.e. the face values, the number of shares, or equivalent) of the assets. Second, the covariance matrix Ω is endogenously determined either if we let the covariance matrix $\tilde{\Omega}$ of the end prices $\tilde{P}_i$ be endogenously affected by debt management, or if we let today's prices P_i be endogenous, or both. The decomposition of Ω into the two parts $1/P_iP_j$ and $\tilde{\Omega}$ spells out these two approaches clearly, and in the analysis below we will assume $\tilde{\Omega}$ to be constant while letting P_i and P_j be endogenously determined.

Third, the expression on the right-hand side is the investor's wealth $W \equiv P_1\bar{q}_1 + P_2\bar{q}_2 + \ldots + P_n\bar{q}_n$, defined in value terms. The investor's initial endowments $\bar{q}_i$ are necessarily identical to supplies q_i^s at an initial equilibrium. When the government engages in debt management, however, it becomes important to distinguish between $\bar{q}_i$ and q_i^s. A debt management operation can be done either in the form of helicopter drops of assets, or in the form of regular open-market operations. In the former case $dq_i^s = d\bar{q}_i$ and the comparative-static derivative $\partial r_i^e/\partial \bar{q}_i$ tells us how asset yields are affected by that kind of debt management. In the following, however, we will use the latter approach and assume that the government buys and sells bonds in the market. Thus the q_i^s will change, but the $\bar{q}_i$ will remain constant, and the derivative $\partial r_i^e/\partial q_i^s$ will in general be different from $\partial r_i^e/\partial \bar{q}_i$.

Taking the system (16)–(17) as a point of departure, we can see that the analysis in previous sections implicitly assumed that

(i) changes in yields r_i^c took the form of changes in 'tomorrow's' prices $\tilde{P}_i^c$, while today's prices P_i were assumed to remain constant;

(ii) the covariance matrix Ω was assumed to be constant.

With this set of assumptions, which is the standard one implicitly underlying all empirical work on debt management, B, Π, and W in (16) will remain constant throughout the analysis, and the only equilibrating changes will take place in the r^e vector. Also, the supply vector α^s will remain constant apart from the jth and ith element in which the debt management operation takes place.

When current price changes are accounted for, there are three induced effects in the system (16):

70 Jonas Agell and Mats Persson

(i) *Wealth effects* A debt management experiment will make asset prices change so that wealth W is changed.[3]

(ii) *Covariance effects* The covariance matrix Ω is endogenously determined, either because both (P_i, P_j) and $\tilde{\Omega}$ are affected by the experiment, or in the simpler case because at least (P_i, P_j) are affected.

(iii) *Supply effects* Writing the supply in value terms like in (16) decomposes it into a physical supply and a price. A change in the physical supplies dq_i^s and dq_j^s will change all prices[4] $P_1, \ldots, P_n$, thereby affecting all elements in the supply vector in value terms and not just two elements of it.

6.2 Debt Management when Asset Prices are Endogenous

To highlight the basic mechanisms involved, we will in the following use a special version of the three-asset model used in Chapter 4. Abstracting from inflation uncertainty, short-term debt instruments become riskless.[5] Then the equilibrium system (16) reduces to

$$
\begin{bmatrix} P_1 q_1^s \\ P_2 q_2^s \end{bmatrix} = \frac{1}{c} \begin{bmatrix} \dfrac{\tilde{\sigma}_{11}}{P_1^2} & \dfrac{\tilde{\sigma}_{12}}{P_1 P_2} \\[2ex] \dfrac{\tilde{\sigma}_{12}}{P_1 P_2} & \dfrac{\tilde{\sigma}_{22}}{P_2^2} \end{bmatrix}^{-1} \begin{bmatrix} \dfrac{\tilde{P}_1^c - P_1}{P_1} - r_3 \\[2ex] \dfrac{\tilde{P}_2^c - P_2}{P_2} - r_3 \end{bmatrix} (P_1 \bar{q}_1 + P_2 \bar{q}_2 + P_3 \bar{q}_3), \quad (18)
$$

where P_1, P_2, and P_3 are the current prices of corporate equity, long-term debt, and short-term debt, and $\tilde{\sigma}_{12}$ is the covariance of

---

[3] One would like to think that, since the assumption of constant relative risk aversion makes the asset demand functions (16) linearly homogeneous in wealth, possible wealth effects would not matter for the equilibrium solution. This is however not true. First, if investors are heterogeneous (for example in terms of risk aversion, subjective perception of the covariance matrix, or differential taxation), wealth effects will affect people differently, thereby affecting the equilibrium solution. Second, even with one representative investor, wealth effects will matter. Since one asset is treated as a numeraire, its price will not increase if wealth effects make demand increase. The equilibrium prices of all other assets will however go up, which implies that the composition of asset supplies in value terms must change and that the expected equilibrium returns adjust.

[4] Apart from the numeraire.

[5] This assumption is perhaps not too unrealistic; as noted in Sect. 3.2 above, the estimated conditional variance of the real yield on short-term debt is a very small number.

Does Debt Management Matter? 71

the end prices $\tilde{P}_1$ and $\tilde{P}_2$. As before, our asset market model is only capable of determining a set of relative rates of return. Consequently, we take the return on short-term debt r_3 as given and treat the expected absolute returns $(\tilde{P}_i^c - P_i)/P_i$ on equity and long-term debt as endogenous. Also, since short-term debt is our numeraire asset, we can without loss of generalization set P_3 equal to unity.

Now, (18) is the basic equation system used to infer the effects of government debt management. To provide a clear-cut benchmark for the ensuing analysis, we first examine the case when valuation changes do not matter. This corresponds to the standard procedure used in the literature, implying the determination of expected asset returns given the assumption that current asset prices are fixed. As already noted, this case is formally equivalent to viewing (18) as an equilibrium system determining the expected end prices $\tilde{P}_1^c$ and $\tilde{P}_2^c$ for given values of P_1 and P_2. We assume that the government performs an open-market operation of the form $P_2 dq_2^s = -dq_3^s$, implying the substitution of dq_2^s units of long-term bonds for $-P_2 dq_2^s$ units of short-term bonds. This experiment, implying that the dollar value of long-term bonds sold equals the dollar value of short-term bonds bought, is carried out by totally differentiating (18) with respect to q_2^s, $\tilde{P}_1^c$, and $\tilde{P}_2^c$.[6] The solution turns out to be

$$\frac{\partial \tilde{P}_1^c}{\partial q_2^s} = \frac{c}{W} \tilde{\sigma}_{12}$$

$$\frac{\partial \tilde{P}_2^c}{\partial q_2^s} = \frac{c}{W} \tilde{\sigma}_{22}.$$

Normalizing by setting $P_i = 1$ and using the definition $r_i^c \equiv (\tilde{P}_i^c - 1)$, these comparative-static results obviously imply that

$$\frac{\partial r_1^c}{\partial q_2^s} = \frac{c}{W} \tilde{\sigma}_{12} \qquad (19)$$

$$\frac{\partial r_2^c}{\partial q_2^s} = \frac{c}{W} \tilde{\sigma}_{22}. \qquad (20)$$

[6] Since we assume that the initial endowments q_i are unchanged, this comparative–static experiment represents a regular open-market operation where the increased supply of long-term debt is balanced by a reduced supply of $P_2 dq_2^s$ bond units in the (redundant) market for short-term debt.

72   *Jonas Agell and Mats Persson*

Equations (19) and (20) conform with well-established intuition. An increase in the supply of long-term debt in exchange for short-term debt drives up the expected long-term bond yield. The response of the expected equity yield depends on the sign of $\tilde{\sigma}_{12}$. When $\tilde{\sigma}_{12}$ is positive, long-term debt and equity are substitutes[7] and lengthening the maturity composition of outstanding debt increases the expected equity yield. In the case of complementarity, $\tilde{\sigma}_{12}$ is negative and our debt management experiment lowers the expected equity yield; i.e., we obtain a 'crowding-in' effect.

We will next see how these results are affected when we drop the assumption that today's prices P_i are fixed. We thus examine the polar case when the probability distribution of end prices $\tilde{P}_i$ is given and current asset prices are endogenous. As a by-product of this, apart from introducing wealth effects and making the supply vector endogenous in value terms, we will also be able to treat the covariance of asset yields $\mathrm{cov}(r_i, r_j) \equiv \tilde{\sigma}_{ij}/P_i P_j$ as endogenous, although we still assume that the covariance of future prices $\tilde{\sigma}_{ij}$ is exogenously given.

As before, we examine a regular open-market sale of long-term bonds, satisfying $P_2 dq_2^s = -dq_3^s$. With exogenous current prices, this flow-based definition of debt management is identical to the stock-based definition saying that debt management operations are experiments leaving the total value of government debt unchanged. The equivalence between the stock and flow definitions of debt management, where one implies the other, does not hold with endogenous current asset prices. When P_2 is endogenous, our flow constraint $P_2 dq_2^s = -dq_3^s$ will thus generally go hand in hand with revaluations of the outstanding stock of government debt.[8]

---

[7] Only in the case of just two risky assets is there equivalence between the sign of σ_{ij} and propositions concerning whether assets i and j are complements or substitutes. See Blanchard and Plantes (1977).

[8] In Chs. 3 and 4 above, where today's prices were constant, the flow definition of debt management $P_2 dq_2^s = dq_3^s$ also implies that the market value of outstanding government debt, $B = P_2 q_2^s + q_3^s$, is constant. In this chapter, where prices are endogenous, keeping B constant would require a trading rule according to

$$P_2 dq_2^s = -q_2^s dP_2 - dq_3^s, \tag{i}$$

which differs by the term $-q_2^s dP_2$ from our flow definition of debt management. The stock trading rule (i) has previously been used by Roley (1979) and by Agell and Persson (1987) as *the* definition of debt management in models with endogenous prices.

Does Debt Management Matter? 73

Totally differentiating (18) with respect to q_2^s, P_1, and P_2 and evaluating all partial derivatives at an initial equilibrium where $P_1 = P_2 = 1$, $q_2^s = \bar{q}_2$, and $q_3^s = \bar{q}_3$, we obtain the general equilibrium partial derivatives

$$\frac{\partial r_1^e}{\partial q_2^s} = \frac{c\tilde{P}_1^e[\tilde{\sigma}_{12} + c\bar{\alpha}_1\bar{\alpha}_2(\tilde{\sigma}_{11}\tilde{\sigma}_{22} - \tilde{\sigma}_{12}^2)/(1 + r_3)]}{W[1 + r_3 - c\,\mathrm{var}(r_m)]} \tag{21}$$

$$\frac{\partial r_2^e}{\partial q_2^s} = \frac{c\tilde{P}_2^e[\tilde{\sigma}_{22} - c\bar{\alpha}_1^2(\tilde{\sigma}_{11}\tilde{\sigma}_{22} - \tilde{\sigma}_{12}^2)/(1 + r_3)]}{W[1 + r_3 - c\,\mathrm{var}(r_m)]}, \tag{22}$$

where we have used the definition $r_i^e = (\tilde{P}_i^e - P_i)/P_i$, W is the initial equilibrium value of initial wealth, $\bar{\alpha}_i$ is the initial portfolio share $\bar{q}_i/(\bar{q}_1 + \bar{q}_2 + \bar{q}_3)$ of asset i, and

$$\mathrm{var}(r_m) \equiv \bar{\alpha}_1^2\tilde{\sigma}_{11} + \bar{\alpha}_2^2\tilde{\sigma}_{22} + 2\bar{\alpha}_1\bar{\alpha}_2\tilde{\sigma}_{12}$$

is the return variance of the market portfolio at the initial equilibrium.

To make some headway in interpreting (21) and (22), we first need some additional information. With constant relative risk aversion, the first-order conditions for individual portfolio optimum in the initial equilibrium are[9]

$$\tilde{P}_1^e - 1 - r_3 = c(\bar{\alpha}_1\tilde{\sigma}_{11} + \bar{\alpha}_2\tilde{\sigma}_{12}) \tag{23}$$

$$\tilde{P}_2^e - 1 - r_3 = c(\bar{\alpha}_1\tilde{\sigma}_{12} + \bar{\alpha}_2\tilde{\sigma}_{22}). \tag{24}$$

Using the expressions for $\tilde{P}_i^e$ implicit in (23) and (24), it is easily seen that, in an economy where the initial supplies of equity and long-term debt are zero ($\bar{q}_1 = \bar{q}_2 = 0 = \bar{\alpha}_1 = \bar{\alpha}_2$), price effects do not matter and (21) and (22) reduce to (19) and (20). Also, local stability of the equilibrium system (18) around some equilibrium point (P_1^*, P_2^*) requires that the denominator in (21) and (22) is positive.[10]

---

[9] Equations (23) and (24) follow from the optimization problem

$$\max_{\alpha_1, \alpha_2, \alpha_3} \ U = \sum_{i=1}^{3} \alpha_i r_i^e - \frac{c}{2}\sum_{i=1}^{2}\sum_{j=1}^{2} \alpha_i\alpha_j\tilde{\sigma}_{ij}$$

subject to $\alpha_1 + \alpha_2 + \alpha_3 = 1$.

[10] Invoking a Walrasian price adjustment rule and taking a first-order Taylor expansion of (18) around an equilibrium point $(P_1^*, P_2^*) = (1, 1)$ yields the system

$$\begin{bmatrix} dP_1/dt \\ dP_2/dt \end{bmatrix} = h \begin{bmatrix} a_{11} & a_{12} \\ a_{21} & a_{22} \end{bmatrix} \begin{bmatrix} P_1 - 1 \\ P_2 - 1 \end{bmatrix} \tag{ii}$$

74   *Jonas Agell and Mats Persson*

Adopting this sign convention, several observations are in order. First, it is trivially true that the quantitative effects with endogenous current asset prices will differ from those obtained with exogenous prices. Thus, the induced wealth, supply, and covariance effects will generally make both the numerator and denominator of (21) and (22) differ from those of (19) and (20). However, with one important exception, the qualitative results prove to be robust. Turning to the basic crowding-out issue in the economics of debt management, we see that the second term in the numerator in (21) is always non-negative.[11] Thus, whenever $\tilde{\sigma}_{12}$ is positive, signalling that long-term debt and corporate equity are substitutes, we still obtain the result that lengthening the maturity composition of government debt leads to crowding-out by increasing the expected equity yield.

However, in the case when $\tilde{\sigma}_{12}$ is negative, we previously derived an unambiguous crowding-in effect. With endogenous prices this is no longer true. Depending on the configuration of initial portfolio holdings, relative risk aversion, and covariances, the numerator in (21) may well be positive, implying a crowding-out effect also in the case when long-term debt and equity are complements.

In sum, we have thus found that allowing for endogenous price changes also serves to strengthen the crowding-out case associated with long-term debt financing in much of the literature. Whether the effects will be quantitatively larger or smaller than those represented by (19) is however still an open question. Depending on the precise values of the additional terms entering the numerator and denominator of (21), compared with (19), allowing for valuation changes may thus serve to magnify or diminish the effects obtained in previous literature.

where h $\equiv$ a positive adjustment rate coefficient

$$a_{11} \equiv \bar{\alpha}_1(\tilde{\sigma}_{22}\hat{r}_1 - \tilde{\sigma}_{12}\hat{r}_2) - \tilde{\sigma}_{22}(1 + r_3)$$
$$a_{12} \equiv \bar{\alpha}_2(\tilde{\sigma}_{22}\hat{r}_1 - \tilde{\sigma}_{12}\hat{r}_2) + \tilde{\sigma}_{12}(1 + r_3)$$
$$a_{21} \equiv \bar{\alpha}_1(\tilde{\sigma}_{11}\hat{r}_2 - \tilde{\sigma}_{12}\hat{r}_1) + \tilde{\sigma}_{12}(1 + r_3)$$
$$a_{22} \equiv \bar{\alpha}_2(\tilde{\sigma}_{11}\hat{r}_2 - \tilde{\sigma}_{12}\hat{r}_1) - \tilde{\sigma}_{11}(1 + r_3)$$
$$\hat{r}_i \equiv \tilde{P}_i^e - 1 - r_3.$$

The system (ii) is locally stable if $a_{11} + a_{22} < 0$, and $a_{11}a_{22} - a_{12}a_{21} > 0$. It is straightforward to show that the latter inequality reduces to the condition that the denominator in (21) and (22) is positive.

[11] This is so because $\tilde{\sigma}_{11}\tilde{\sigma}_{22} - \tilde{\sigma}_{12}^2 \geq 0$.

Does Debt Management Matter? 75

Finally, turning to the own-yield effect of an open-market sale of long-term debt, it may at a first glance appear as if the numerator in (22) can become negative, thus implying a decrease in the expected yield on long-term debt. However, by the sign restriction imposed on the denominator, we can always rule out this paradoxical result;[12] also, with endogenous prices the own-yield effect must be positive. As before, however, we cannot determine a priori whether allowing for price changes magnifies the original results.

6.3 The Empirical Evidence

In principle, it is easy to construct hypothetical numerical examples where allowing for return adjustments through variations in current asset prices may make a significant difference to the results. From an empirical point of view, however, it is of greater interest to examine whether the policy derivatives in (21) and (22) differ from those in (19) and (20) when using real-world data on asset returns and portfolio holdings. For this reason, we have computed the policy derivatives using US quarterly data for 1970(1)–1988(2).

As the policy derivatives in (21) and (22) contain some parameters not included in previous sections, we have made some adjustments to the original data set. In accordance with the derivation of (21) and (22), we assume an initial equilibrium in each quarter where $P_1 = P_2 = 1$. The expected end-of-quarter prices $\tilde{P}_1^e$ and $\tilde{P}_2^e$ are then calculated as

$$\tilde{P}_i^e = 1 + d_i + \hat{g}_i,$$

where d_i is the known dividend yield (i.e. dividends on common stock and coupon on long-term bonds) and $\hat{g}_i$ is the predicted nominal capital gain obtained from the moving-sample vector autoregression procedures discussed above.

Since the initial equilibrium prices are normalized to unity,

---

[12] The numerator in (22) is positive whenever

$$1 + r_3 - c\bar{\alpha}_1^2 \tilde{\sigma}_{11}(1 - \rho_{12}^2) > 0. \tag{iii}$$

However, since $\mathrm{var}(r_m) > \bar{\alpha}_1^2 \tilde{\sigma}_{11}(1 - \rho_{12}^2)$, the sign restriction on the denominator implies that (iii) holds.

76   *Jonas Agell and Mats Persson*

the covariances of end-of-period asset prices are assumed to be identical to the conditional covariances of nominal asset yields implied by the VAR model.[13] The initial portfolio shares represent the actual composition of the aggregate financial wealth of US households for each quarter.[14] The empirical counterpart of $\bar{\alpha}_1$ is an aggregate of corporate equity and equity-like assets (e.g. holdings of mutual fund shares). The portfolio share $\bar{\alpha}_2$ corresponds to all other long-term financial assets (corporate and government bonds, tax-exempt state and local bonds) held by households. The residual asset, finally, includes all other liquid and interest-bearing assets. We also normalize the initial equilibrium value W of initial wealth to unity, which implies that the policy derivatives now show the effects of marginally increasing the *fraction* of long-term bonds held by investors.

The calculated policy derivatives for the period 1970(1)–1988(2) are shown in Fig. 6.1. For comparison, the broken curve depicts the results obtained when using the policy derivatives (19) and (20), which apply for the standard model without endogenous price changes. Somewhat surprisingly, and in spite of the theoretical arguments suggesting the importance of allowing for the endogeneity of asset prices, the results obtained in the case of endogenous asset prices corresponds closely to those derived in the case of exogenous prices.

To conclude, these results suggest that allowing for valuation changes does not in a significant way alter the yield effects obtained using the standard setup (discussed in Chapter 4), where current asset prices are implicitly treated as exogenous. However, since the effects of debt management ultimately depend on the interaction of the real and financial sectors of the economy, incorporating endogenous asset prices may still be warranted. In macro models where current asset prices are taken as given, the transmission mechanism linking relative asset supplies to the spending decisions of firms and households is restricted to the relative costs of financing the spending in question. In models allowing for endogenous asset prices, there will be an additional transmission channel arising from the wealth effects implied by valuation changes on outstanding assets.

---

[13] By definition, $\text{cov}(r_1, r_2) \equiv \text{cov}\,[(\tilde{P}_1 - P_1)/P_1, (\tilde{P}_2 - P_2)/P_2]$. When $P_1 = P_2 = 1$, this reduces to $\text{cov}(r_1, r_2) \equiv \text{cov}(\tilde{P}_1, \tilde{P}_2)$.

[14] The data were kindly provided by Prof. Benjamin M. Friedman.

Does Debt Management Matter?

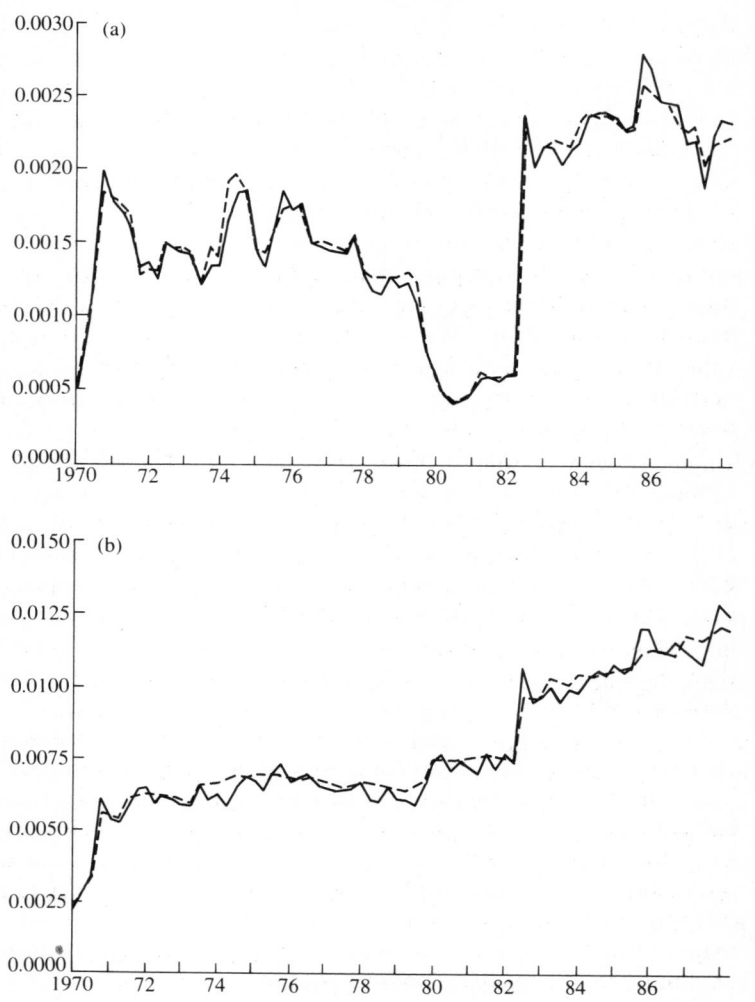

FIG. 6.1 (a) The derivative $\partial r_1^c / \partial \alpha_2^s$ for models with and without valuation changes: quarterly data 1970(1)–1988(2)
(Solid curve: model with valuation changes)
(b) The derivative $\partial r_2^c / \partial \alpha_2^s$ for models with and without valuation changes: quarterly data, 1970(1)–1988(2)
(Solid curve: model with valuation changes)

7

Summary and Conclusions

The preceding chapters have analysed the effects of government debt management. For analytical simplicity, we have been careful to define debt management as involving open-market operations only; i.e., for given government expenditures and taxes we have considered the substitution of, for example, short-term for long-term government bonds. After reviewing the literature and bringing in some evidence of our own, several points stand out.

First, most empirical work on debt management—including the empirical parts of the present paper—focuses on the effects of the maturity composition of government debt on relative asset yields in general, and on corporate equity yields in particular. In the real world, however, there is no such simple and single-dimensional characterization of debt management. Recognizing the wide array of debt instruments available to the government (e.g. tax-exempt versus taxable bonds, marketable versus non-marketable assets, etc.) and the multitude of conceivable policy targets unrelated to the crowding-in/crowding-out issue, the design of debt policy entails a choice between a continuum of different debt attributes to attain many different goals. As a consequence, discussing the stance of debt management only in terms of the maturity composition of government debt is potentially misleading.

Second, using the mean–variance 'work-horse' model of portfolio choice and asset market equilibrium, we turned to the empirical evidence. Upon invoking a moving-sample vector autoregression model to allow for time-varying risk perceptions, we examined the effects on relative asset yields of lengthening the maturity composition of government debt. It turned out that these effects were rather small in magnitude, and that their numerial values were highly volatile. Thus, the policy conclusion to be drawn seems to be that there is not much scope for a debt management policy aimed at systematically affecting asset yields.

Does Debt Management Matter? 79

This conclusion is further strengthened if we take into account the possibility of Ricardian equivalence discussed in Section 2.2 above.

Third, we examined whether the results are robust with respect to the choice of time unit implicitly made when constructing the data. Rerunning the model with monthly instead of quarterly data, we found that the qualitative conclusions remain intact: increasing the supply of long-term bonds in exchange for short-term bonds still tends to raise the expected yields on equity and long-term bonds. In quantitative terms, however, the yield responses now seemed stronger. This suggests that the frequently neglected time aggregation problem should be of more concern in future empirical work.

Fourth, since the above results were based on computing conditional covariance matrices based on the information contained in historical return data, we have tested the robustness of the policy conclusions by using an alternative data source, namely the implicit variances obtained from options data. It turned out that this data set produced results that differed substantially from those obtained using a more conventional vector autoregression procedure: the yield responses to changing relative asset supplies were both considerably smaller and more volatile over time than those obtained using historical return data.

Fifth, we have dropped the standard maintained assumption of constant current asset prices, thereby allowing for more complex policy effects than in previous studies. It turns out that the model is surprisingly robust to this. Allowing for endogenous asset prices—which in turn implies various effects on wealth, on asset supplies (in value terms), and on covariances—hardly affects the results.

These are several problems associated with the present set-up. For example, a satisfactory dynamic formulation of the model (allowing—among other things—for a consistent treatment of term structure problems) is still non-existent. In previous empirical work, today's prices have been more or less implicitly assumed to be constant and unaffected by debt management, while asset yields have adjusted via changes in tomorrow's prices. And since the model is strictly atemporal, there is no date after tomorrow. In Chapter 6 we tried the alternative of assuming (the probability distribution of) tomorrow's prices remaining unaffected by debt management, while today's prices change instead. In reality, of course, both prices are affected—as are the prices at dates after tomorrow. This however calls for a fully dynamic model, which consistently integrates the real and financial sides of the economy. The problems associated with constructing such a model provide a promising research agenda for the future (cf. Cox *et al*. 1985).

Another suggested line of research is due to the fact that the supplies of financial assets are treated in an overly simplistic manner. The portfolio model outlined in Chapter 3 assumed an exogenous net supply of assets, including both the government's and the private firms' supplies of bonds and shares. We know, however, that firms' financial decisions are more complex than that, and thus a more realistic representation of corporate financial decisions is warranted.

On a less ambitious level, there are several interesting topics for further research. For example, we have here dealt with a very limited menu of assets. It would be straightforward to include a few more, e.g. real estate and consumer durables. This would illuminate how sensitive the policy conclusions are to the choice of assets included. However, real estate and consumer durables would place some restrictions on the choice of the data interval, since reliable yield figures are not available for short time intervals.

Still, the most important item in most household portfolios is missing, namely human capital. The analysis of non-marketable assets is fairly straightforward in principle (cf. Mayers 1972), but in practice we will suffer from the lack of yield data. As for other types of assets with a limited degree of marketability, for example pensions savings, data are however available. Such assets also point to the need for studying a model with heterogeneous investors, for example insurance companies and pension funds as well as non-financial companies. The inclusion of such investors requires us to take into account the specific tax situation of each investor category, as well as the fact that such investors are ultimately owned by the households,[1] who constitute the basic agents in the markets.

---

[1] This is to some extent done in Agell and Persson (1987).

8

Comment

JEFFREY A. FRANKEL

The topic of debt management and its possible effects on the stock of real capital via portfolio crowding-out has been relatively overshadowed in recent years by the more fundamental question of whether debt might be altogether neutral in its effects. But it is still an important topic, and increasingly so in countries that are smaller than the USA and are only now developing free and open markets in debt. That the literature is of a manageable size is certainly an advantage. The above study by Agell and Persson covers all the relevant issues nicely and makes a good start at trying out the various alternative econometric approaches with which one might address the question.

The portfolio balance framework, which as the authors say is the 'work-horse' model used to address such questions, is easily stated. In terms of their equation (5),

$$\alpha = \pi + \beta r^e. \tag{1}$$

where α represents the shares of the investor's portfolio that he or she allocates to various assets, and r^e represents the expected rates of return on the various assets. (Where Agell and Persson have a coefficient $1/c$ in front of the matrix of substitutability B, I have here subsumed the coefficient into β.) The question of whether a change in the maturity composition of debt makes a difference for required rates of return on short- and long-term debt is the question of how closely substitutable are the two kinds of debt in investors' portfolios. If the two assets are perfect substitutes, i.e. if the relevant element of β is infinite, then the

The author would like to thank Jonas Agell and Mats Persson for the opportunity to examine their research, and Villy Bergstrom and FIEF for the opportunity to examine Stockholm.

82 *Jeffrey A. Frankel*

composition of the debt has no effect. For the effect on the required rate of return on real capital, the question is, how closely substitutable are long-term debt and real capital? Effects on the rates of return can be seen more clearly by inverting the equation:

$$r^e = -\beta^{-1}\pi + \beta^{-1}\alpha. \tag{2}$$

(This is equation (9) in Agell and Persson above.) Now it is clear that, if the relevant element of β is infinite, the corresponding effect on the rates of return will be zero.

There is more than one general approach that can be taken to estimating such a system of equations econometrically.

8.1 Regression of Asset Shares and Expected Returns

The first variety of the regression approach consists of regressing asset shares α against returns, as in equation (1), to estimate the matrix of substitutability coefficients.[1] This may be the better way if one thinks that errors in the asset demand functions (factors omitted from the simple portfolio balance theory) are large. But a major drawback is the large errors one makes in using the observable *ex post* returns r in place of the correct, but unobservable, *ex ante* expectations of investors.

The alternative is to run the regression in inverted form, as in equation (2).[2] There are two advantages to doing it this way. First one can use *ex post* returns r to measure expectations r^e, under the rational expectations argument that

$$r = r^e + \varepsilon,$$

where ε, the forecasting error made by investors, is assumed to be purely random. Substituting into equation (2) gives us

$$r = -\beta^{-1}\pi + \beta^{-1}\alpha + \varepsilon. \tag{3}$$

This is a suitable equation on which to run a regression, because the rational expectations assumption says that the error term is uncorrelated with all information available at time t, including α.

---

[1] This is the approach followed by Smith and Brainard (1976), Backus and Purvis (1980), Backus *et al.* (1980), and Friedman (1978, 1985*b*).

[2] Example of regressions in inverted form include Fair and Malkiel (1971) and Modigliani and Sutch (1966, 1967).

Does Debt Management Matter? 83

The second advantage is that the inverted system can be thought of as the market equilibrium condition. It tells us what the rate of return has to be for given asset supplies to be willingly held. In particular, perfect substitutability between two assets holds when the relevant element of β is infinite, which is more testable in inverted form: it is simply the condition that the relevant row of β^{-1} consists of zeros.

The best method may be to use simultaneous equation estimation to allow for both kinds of errors, i.e. omitted factors in the portfolio balance equation and the expectational errors that investors make in predicting returns.[3] (The difficulty, as so often, is finding proper instrumental variables.) But however the equation is estimated, the results have always been very imprecise, and often implausible in sign or magnitude. Such difficulties induce one to look around for additional information that can be brought to bear on the subject.[4] A promising source of additional information is the theory of optimal portfolio diversification, which leads us to the second way in which one can go about estimating the parameters.

8.2 Estimation of the Optimally Diversified Portfolio

If investors choose their asset shares, α, optimally to maximize a function of the mean and variance of their wealth, then the parameters of the asset demand equation are bound by the simple constraint

$$\beta^{-1} = c\Omega, \tag{4}$$

where c is the coefficient of relative risk aversion and Ω is the variance–covariance matrix of real returns on the various assets. We can now see that the case when assets are highly substitutable is the case when risk is not very important, either because uncertainty is low (the elements of Ω are small) or because investors are not very risk-averse (c is small).

The traditional way of estimating Ω is to compute the variance–covariance matrix of returns over a particular period of

---

[3] Citations include Friedman (1977), Roley (1982), and Masson (1978).

[4] For example, many of the authors cited in fn. 1 have thought it necessary to impose *a priori* some element of their own beliefs on the estimates.

84   *Jeffrey A. Frankel*

time. This is the approach that Agell and Persson start with (Section 4.1).[5] As they recognize, it has several problems. First, it does not facilitate the testing of the proposition that investors do in fact choose their portfolios optimally, which one would like to be able to do before imposing the proposition. Second, the technique of computing Ω from observed squared deviations of the rate of return around its sample mean does not allow expected returns to vary over time; thus, it is inconsistent with the sort of policy experiment one is trying to answer, the effect of a change in the composition of assets on the expected rate of return.

A relatively simple way to allow the expected returns to vary over time is to estimate a vector autoregressive (VAR) process on the returns, as Agell and Persson do (Section 4.2).[6] Better yet, one can allow the coefficients to change over time with a 'moving sample', as in the technique of 'rolling regressions', which the authors also try (Section 4.3).[7] Although these approaches are a major improvement over the traditional way of estimating Ω, they are based on the assumption that this period's expected returns are determined by last period's returns. Thus, they still do not lend themselves to experiments where expected returns change suddenly because of a contemporaneous change in monetary or debt management policy, or to news about future monetary or debt management policy.

8.3   The Constrained Asset Share Estimation Method

Several years ago, I proposed a technique for estimating asset demand functions which I believe combines the best of the previous two approaches (Frankel 1982, 1985; Engel *et al.* 1989). Substitute equation (4) into equation (3):

$$r = -\beta^{-1}\pi + (c\Omega)\alpha + \varepsilon. \tag{5}$$

[5] Examples of this approach include Roley (1979) and the studies by Ben Friedman.

[6] One example of this approach is Bodie *et al.* (1983), who estimate expected returns as a univariate autoregressive process.

[7] The technique has also been applied by Friedman (1985*b*, 1986). Friedman and Kuttner (1988) give declining weight to observations with longer lags.

Does Debt Management Matter? 85

Ω is the variance–covariance matrix of *ex post* returns as perceived by investors; that is, it is the variance–covariance term of their forecasting errors ε. In other terms, $\Omega = \text{var}(\varepsilon)$. Thus, the technique calls for estimating the equation subject to the constraint that the coefficient on the asset shares is proportional to the variance–covariance matrix of the error term. Estimating a system of equations subject to a constraint between the coefficient matrix and the error variance–covariance matrix is unusual in econometrics (as compared to constraints among the coefficients). But it can be done by nonlinear maximum likelihood estimation. We now call this the constrained asset share estimation (or CASE) method. The advantages of the technique are, first, that one can test the optimization hypothesis (simply by a likelihood ratio test on the constraint (4)), and second, that it allows the expected returns to vary freely.

The application of the CASE method still leaves one with several basic questions. The first is whether the assumption of optimal portfolio diversification is in fact justified. Second is the objection that, if one is to make such a fuss about letting the expectations r^e vary over time (the first moments of the conditional distribution of asset returns), then should one not also allow the variance–covariance matrix Ω to vary over time (the *second* moments)? Third is the question—which applies as much to the regression method (equation (3) above) as to the CASE method—of whether the treatment of expectations as forward-looking is correct (i.e. whether one can infer what investors believe *ex ante* from what is observed to happen in a given sample period *ex post*). I will take the three in reverse order.

I am quite sympathetic to the third point. The rational expectations methodology is absolutely standard in the modern macroeconomic literature, of course. In both the literature on the foreign exchange risk premium and the literature on the term structure of interest rates, for example, the methodology of inferring what investors must have expected from what is observed to happen *ex post* is the norm. Yet it can yield strange conclusions. Studies using survey data on the expectations of market participants give very different results from the rational expectations methodology. The survey data results are more in keeping with simple notions that expectations are in part reflected in the forward discount (in the case of the foreign exchange market) or

86   Jeffrey A. Frankel

in the term structure of interest rates.[8] For better or worse, however, the rational expectations methodology continues to reign supreme.

The second point, which some have cited as a limitation of my technique, is not really a problem at all. One can allow the variance–covariance matrix Ω to change over time by using the auto-regressive conditional heteroskedasticity (ARCH) method of Rob Engel, while maintaining the constraint of the CASE method (Engel and Rodriguez 1989; Bollerslev *et al.* 1988; Engel *et al.* 1989). While the evidence does support the need to let Ω move over time, the results tend in other regards to be similar to results obtained in the earlier studies, which constrained Ω to be constant.

Another way that one can deal with the need to let the variances change over time is to extract from options prices estimates of what the market thinks the variances are. Assuming that the theoretical option-pricing equation that is used to extract the variances is correct, the technique has a major advantage. It allows the investors' subjective variances to vary freely, not just in the slow-moving manner of ARCH. Agell and Persson do look at options prices (Chapter 5), and they find that the implicit variances are generally much smaller than the variances measured from *ex post* data, but move around more.

The remaining problem with using the CASE method may be thornier, depending partly on one's a priori degree of attachment to the optimizing paradigm. The data appear to be at war with the hypothesis of optimal portfolio diversification. If the coefficient of relative risk aversion c is not constrained then the estimates can be quite plausible. Consider the example of recent US fiscal deficits in the area of $150 billion. They represent yearly issues of debt of roughly 1 per cent of wealth, where we are including corporate assets and real estate in wealth, but not human wealth or the wealth of non-US residents. Some estimates in Frankel (1985) say that such an increase in Treasury debt drives up the expected relative rate of return on Treasury debt by 0.462 per cent, which, though low, is not completely unreason-

---

[8] References are Froot and Frankel (1989) and Froot (1989), respectively. Friedman (1986) uses survey data on long-term interest rates, stock prices, and inflation rates to help estimate directly investors' reactions to expected rates of return.

Does Debt Management Matter? 87

able. But this is for c estimated at 110.3, which is much higher than conventional estimates of the coefficient of risk aversion like 2 or 4. If c is constrained to 2, then the estimated effect of a debt increase equal to 1 per cent of wealth is a very small 0.008 per cent—less than one basis point!

A related difficulty is that tests of the optimization hypothesis in fact reject the constraint. Thus, using the technique, or any other variant of optimal portfolio diversification, requires that one accept the optimization assumption on a priori grounds.

The problem lies not in the estimation technique, but in the inherent nature of the hypothesis of optimal portfolio diversification. In my estimates, the variances of per annum returns on stocks and bonds are in the vicinity of 0.004. Readers are invited to choose their own favourite estimate of the coefficient of relative risk aversion c, and multiply it against 0.004 as in equation (4), to see how small the elements of β^{-1} must be. Agell and Persson (and Friedman before them) get roughly similar variances. For the case of long-term bonds, the unconditional variance of the per annum return is estimated at 0.0145 per cent ($= 4 \times 0.0036$ in Table 4.2(a)). The conditional variance is slightly smaller than this (owing to the 'shrinkage' of taking out the forecastable component): 0.012 per cent ($= 4 \times 0.003$ in Table 4.3(b)). These estimates of the variance are somewhat higher than mine, perhaps for the interesting reasons relating to frequency of observation that Agell and Persson discuss. So they imply an effect of the debt supply on the rate of return on debt ($2 \times 0.012 = 0.024$ per cent, or $2\frac{1}{2}$ basis points) that is somewhat higher than mine.[9] And my estimate of the effect on the rate of return on equity is *very* small, because the correlation of the returns on long-term debt and equity was close to zero in my sample period (Frankel 1985: 40). But the important point is that *all* these estimates are exceedingly small judged by the standards of policy-making and observed fluctuations in the financial markets. Even under the largest of the estimates, the effects of debt management on rates of return would appear to be too insignificant to bother with.

---

[9] The portfolio effects that Agell and Persson calculate when the variances are estimated from options prices are considerably smaller, in proportion to the variances.

88   *Jeffrey A. Frankel*

At least three reactions to this apparent impasse are possible. The first would be to argue that investors are optimizing in some intertemporal dimension in a more sophisticated manner than is captured in simple mean–variance optimization (which is consistent with maximization of expected intertemporal utility only if either the asset return process is restricted—e.g. to the lognormal—or the utility function is restricted—e.g. to the logarithmic). This point is capable of yielding all manner of theoretical results. The drawback with pursuing it is that it does not lead readily to an alternative approach to empirical estimation (unless one imposes further restrictions, which are typically far more arbitrary than those needed for mean–variance optimization).[10]

The second, quite different, possible reaction is to conclude that investors must be *less* sophisticated than the mean–variance theory supposes. This response would suggest a return to the unconstrained regression approach above. This logic, in effect, is what originally led Agell and Persson to consider what they called a 'revealed preference' approach.

The third possible response to the inconvenient quantitative implications of mean–variance optimization is to push and tug a bit at the parameters to come up with estimates that are not too unreasonable. First, one can assume a coefficient of relative risk aversion greater than 2. Second, one can omit real estate from the definition of wealth. The rationale would be similar to the presumed rationale (beyond questions of data availability) for omitting human wealth and foreign wealth: an argument that only some subset of total wealth is available in portfolios that are relevant for the determination of prices in domestic financial markets. What is gained is that, when the denominator is made smaller, a given increase in government debt like $150 billion becomes a larger fraction α of the portfolio. Thus, it has a larger effect on expected returns in equations (2) or (5). This is the solution that Ben Friedman has settled on. A sympathetic interpretation would argue that it represents a step in the direction of institutional realism, by recognizing that savings that accumulate with institutional investors like pension funds and mutual funds/

---

[10] In any case, Mehra and Prescott (1985), working within the paradigm of intertemporal optimization, come to the same conclusion regarding the implausibly small magnitude of the risk premium.

Does Debt Management Matter? 89

unit trusts have more of an immediate impact on financial markets than the broader total of household wealth. Nevertheless, one should keep in mind that, in theory, real estate, foreign wealth, and human wealth should matter too.

Agell and Persson seem in principle to favour adding real estate, consumer durables, and human wealth back into the analysis (p. 80 above). I would add that, in theory, the portfolios of residents of the rest of the world are equally important, as a result of the increasing integration of international financial markets. The paradigm of high capital mobility applies not just to the largest industrialized countries such as the USA, but increasingly also to smaller countries in Europe and elsewhere. Figure 8.1, taken from Frankel (1989), shows the three-month covered interest differential in the 1980s for 6 out of a set of 25 available countries. The differential is tangible evidence that by 1987 most European countries had open capital markets, including West Germany (and Switzerland), which removed most controls on capital inflows in 1974, the UK, which removed most controls on outflow in 1979, and France and Italy, which retained controls on outflow into the 1980s but have been phasing them out in recent years. The negative differential for Sweden suggests that controls on capital outflow were somewhat binding up to 1985, but that they have not been very important since then.

The final exercise in the paper is to allow for the endogenous effect on current asset prices when computing the effect of changes in supplies on expected returns. The authors are quite correct about statements regarding 'an increase in government debt equal to 1 per cent of wealth': the effect on expected returns will be smaller if one takes into account that a $150 billion increase in the supply of one asset out of a total portfolio of $15,000 billion will have some downward effect on the price of the asset, so that the change in α will be less than 1 per cent. They calculate effects under the simplifying assumption that the expected future price of the entire asset is tied down, so as not to have to specify and solve an intertemporal general equilibrium model. They find that treating the contemporaneous asset prices as endogenous in this way in fact makes little difference for the effects of asset supply changes on expected rates of return.

They express some surprise that endogenizing asset prices

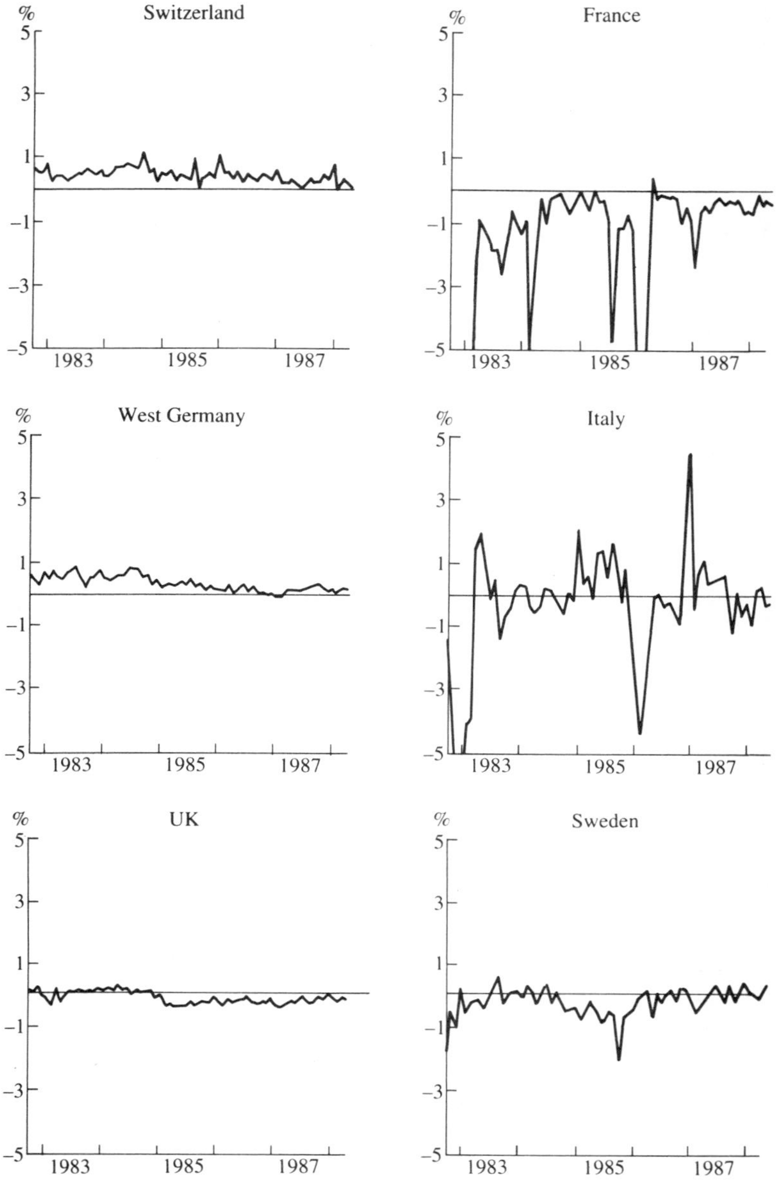

90   *Jeffrey A. Frankel*

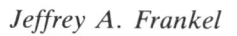

FIG. 8.1   Covered interest differential (3-month local Eurodollar)
(*Source*: Frankel 1989)

Does Debt Management Matter? 91

makes little difference. But it seems to me that this is precisely what one might have predicted. The reasoning is as follows. We saw above that, for ordinary estimates of the magnitude of the variances and of the coefficient of risk aversion, the degree of substitutability among assets is very high. We can express this by saying that it takes only small changes in expected rates of return to have large effects on portfolio shares, or that large changes in portfolio shares have small effects on expected rates of return. The important point here is that this high degree of substitutability prevents large fluctuations in expected future capital gains and therefore ties the current asset price closely to the expected future asset price. Since Agell and Persson assume that the expected future asset prices are unchanged, it is not surprising that the contemporaneous asset prices change very little, from which it follows that the change in α is not very different from what it was when changes in contemporaneous asset prices were ruled out of the experiment.

As I said above, the inconveniently high magnitude of the implied degree of substitutability is not the fault of the researchers, but is inherent in the constraints of optimal portfolio diversification. I tentatively incline to the view that most investors either use a bigger coefficient of risk aversion in their investment decisions than they do in other aspects of their economic and non-economic behaviour, or else measure the risk of acquiring a given asset against a much smaller frame of reference than their entire wealth. Either way, the implication would be that investors act as if they are less ready to make large changes in their holdings in response to small changes in returns than our optimization theory would suggest. In any case, Agell and Persson have provided us with a remarkably comprehensive battery of the alternative existing techniques with which one can estimate these systems.

9

Comment

BENJAMIN M. FRIEDMAN

The paper by Jonas Agell and Mats Persson on debt management policy is an interesting and stimulating contribution for at least two reasons. First, in contrast to the situation that prevailed a generation or so ago, debt management policy is today a much neglected subject. Economists continue to focus intensely on the determination of the total quantity of government liabilities outstanding (fiscal policy) and on the choice of interest- versus non-interest-bearing liabilities within that total (monetary policy), but the choice among different kinds of interest-bearing liabilities— which is the essence of debt management policy—has largely dropped from view. Has this shift of attention occurred because such choices, in fact, have no material effects on financial markets or non-financial economic activity? Or has debt management disappeared from economists' agenda for some other reason? The answer is clearly of interest to practical policy-makers, who are responsible for a country's debt management whether it has economic consequences or not.

More broadly, the above paper by Agell and Persson is of interest because the empirical study of asset demand behaviour, and of the financial market equilibria to which that behaviour gives rise, has only just begun to attract the research effort and attention that it deserves. As most readers will be aware, this subject has already advanced to a fairly high level in theoretical analysis. By contrast, empirical implementation of the central ideas involved in this theoretical work has lagged behind. The paper by Agell and Persson makes a useful contribution along these lines as well.

9.1 Modelling Conditional Variance–Covariance Structures: Treatment of Serial Correlation

Agell and Persson are clearly correct in stating that what matters for asset demand behaviour is the *conditional* variance–covariance structure describing the risks associated with holding various assets—that is, the variance–covariance structure conditional on whatever information investors have at the time they choose their portfolios. Agell and Persson are also correct in maintaining that there is no reason to believe that the unconditional (*ex post*) variance–covariance structure measured over any particular sample period necessarily corresponds to the conditional variance–covariance structure that investors perceive. The main challenge in this line of empirical work, therefore, is to infer the relevant conditional variance–covariance structure—which is, of course, unobservable. This is the principal thrust of Agell and Persson's empirical effort, and it represents the right way to proceed in this research.

I also agree with Agell and Persson (although here there is room for disagreement) that the relevant conditional variance–covariance structure can change over time. Indeed, their work shows substantial changes over time in inferred variance–covariance structures, which they not only show directly (e.g. Figs. 4.2 and 4.3) but also represent in terms of the key derivatives that summarize the implications of debt management actions for equilibrium asset returns (e.g. Fig. 4.4). Even so, the authors do not say *why* the conditional variance–covariance structure might plausibly change over time. The question is an important one, because the treatment of any of a variety of issues that arise in this empirical work might plausibly depend on one's view of just why this structure is changing: Does the world remain unchanged, but people learn its features gradually over time? If so, what determines their learning process? Are some relevant aspects of the stochastic structure of asset returns changing? If so, which aspects, and why are they changing? Are there new institutional structures, or is the nature of the shocks to which the economy in general and the asset markets in particular are subject also changing over time? My sense of the direction in which this line of research is now headed is that empirical

94   *Benjamin M. Friedman*

implementation will increasingly depend on answers to these and similar questions.

The central challenge taken up by Agell and Persson is to represent a variance–covariance structure that is not only unobservable but also (for whatever reason) moving over time. In order to do so, they focus on removing the serial correlation in observed asset returns. It is useful to note explicitly that, for readers who start from the theoretical presumption that there is no serial correlation in the excess returns on risky assets anyway, this procedure will obviously seem puzzling. Nevertheless, by now there is substantial evidence showing that, over just about any sample period one chooses to examine, excess returns on risky assets traded in US markets do in fact exhibit substantial serial correlation.

Removing this serial correlation means that the observed unconditional variances shrink down into smaller conditional variances (and the same typically happens for covariances, measured in absolute value). For example, the conditional variance that Agell and Persson report for the short-term debt asset for the 1960–88 sample period is one-quarter the size of the corresponding unconditional variance (compare Tables 4.1(*a*) and 4.3(*a*)). The reason, of course, is that what makes the real returns on short-term debt instruments stochastic in the first place is uncertain inflation, and the inflation rate, on a quarterly basis, is very highly serially correlated. Hence taking out the serial correlation greatly shrinks down the variance. Agell and Persson report conditional variances for long-term debt and equity returns that are both about three-quarters the size of the corresponding unconditional variances (see again Tables 4.1(*a*) and 4.3(*a*)).[1]

Several years ago, a major problem for empirical work along these lines would have been to reconcile the fact that the empirical heart of such work rests on the presence of serial correlation in the excess returns on risky assets with the theoretical notion that, in an efficient market, there ought not to be such serial correlations. Today, however, there is a growing literature that

[1] Analogous shrinkage factors reported in my work are typically even greater. In Friedman (1985*b*), for example, the ratios of conditional to unconditional variance are roughly one-tenth for short-term debt, one-third for long-term debt, and one-half for equity. Different empirical methodologies can also lead to much smaller shrinkage, however; see Friedman and Kuttner (1992).

Does Debt Management Matter? 95

attempts to explain the presence of serially correlated asset returns on theoretical grounds.[2]

9.2 Temporal Aggregation

As Agell and Persson are clearly aware, any procedure for inferring conditional variance–covariance structures by removing the serial correlation from *ex post* data on asset returns is likely to deliver results that depend crucially on the chosen unit of temporal aggregation. Section 4.4 of their paper focuses in particular on the choice of a quarterly versus monthly time unit.

I agree that temporal aggregation is important in this context, and the results of my own work suggest that the choice of time unit may matter even more than Agell and Persson's results indicate, especially in any context (like this one) in which what principally drives the empirical results is the serial correlation properties of the respective asset return series. For example, in the USA the quarterly excess return on long-term debt exhibits negative first-order serial correlation (see Friedman and Kuttner 1992). On a monthly basis, however, the first-order serial correlation is positive.[3] The reason is that, over most sample periods, observed excess returns on long-term debt instruments are dominated by movements in which the return moves in the same direction for several months, followed by a substantial reversal. (These instances are not frequent, but they are large enough— and there are enough of them—to dominate the time-series properties of the data.) Returns measured from 31 March to 30 June, from 30 June to 30 September, and so on show these instances as single large positive or negative returns, followed by large returns of the opposite sign, so that there is negative serial correlation. By contrast, returns measured from 31 March to 30 April, from 30 April to 31 May, and so on remove the appearance of occasional large, jagged up-and-down movements in the quarterly data, and the corresponding returns are therefore positively serially correlated in the monthly data.

---

[2] See e.g. Poterba and Summers (1987), DeLong *et al.* (1990), and the related literature of 'mean reversion' in asset prices.

[3] The same difference in sign appears in the data for Canada, Japan, and the UK (though not for Germany); see Friedman (1987).

96   Benjamin M. Friedman

Similarly, in other work (see Friedman and Laibson 1989) I
have shown that the observed time-series properties of the excess
return on US equities depends in yet other ways on the time unit
chosen for temporal aggregation. On a quarterly basis, equity
returns exhibit both negative skewness (skewness value -0.61,
significant at the 0.005 level) and leptokurtosis—that is, the
familiar 'fat tails' compared with the normal (kurtosis value 1.16,
significant at the 0.01 level). The same returns, measured on a
monthly basis, are not noticeably negatively skewed (skewness
value -0.08, not significant at any plausible level), but are even
more leptokurtotic (kurtosis value 0.79, significant at the 0.005
level). Clearly, the negative realizations of the unskewed monthly
returns are 'clumping together' in such a way as to produce
negatively skewed quarterly returns. In the context of patterns
such as these, what matters for the determination of market
equilibria is whether the investors that dominate the market are
interested in a one-month horizon or a one-quarter horizon (or
perhaps a daily horizon, or an annual horizon, or even some
other).[4]

Agell and Persson acknowledge this crucial dependence in
their paper. Even so, much research remains to be done—
including research on the nature of broadly defined transaction
costs (always a difficult subject, because of the mathematically
intractable nature of the expressions that arise under even the
most limited generality)—before anyone can say with confidence
what time horizon best matches the typical investor's perception
of asset risk.

9.3   Changes in Conditional Variance–Covariance Structures over Time

In order to allow the conditional variance–covariance structure
to change over time, Agell and Persson model investors' behaviour as if investors pick some previous demarcation point and
consider all experience since that initial point to be equally
relevant to predicting the future, regardless of how far in the past

---

[4] Fischer (1983) and Fischer and Pennacchi (1985) show that the time horizon is
especially important in the context of serially correlated asset returns—which is,
of course, exactly the context of the Agell–Persson paper above.

Does Debt Management Matter? 97

that initial point may be when they form their expectations. This is the same procedure I used in some of my work along these lines (Friedman 1985*b*). It would be the right way to model this aspect of investors' behaviour if the reason the relevant conditional variances and covariances are changing is that a distinct institutional change occurred at some point in the past, since which time the underlying stochastic environment governing asset returns has been stable. In that case, the reason conditional variances and covariances change over time is simply that people are still learning about their new environment, and they do so with a finite sample of observations. If the world undergoes no further changes, eventually the relevant data sample will become infinite, and (in the absence of some pathology) the conditional variance–covariance structure that investors use will converge to the true structure with no future changes. The reflection of this convergence in Agell and Persson's results is that, as their sample period keeps lengthening, unexpected movements of a given size in asset returns make ever less difference for the conditional variance–covariance structure that they infer (see Figs. 4.2 and 4.7).

By contrast, Kenneth Kuttner and I have recently experimented with an alternative procedure, corresponding to the view that institutional changes are not so discreet; instead, large and small institutional changes occur on a continual basis (see Friedman and Kuttner 1992). An investor in 1989 might therefore think that the experience from the previous few years was more relevant than the experience from the 1960s—when, for example, international participation in US markets was negligible compared with current levels. Kuttner and I have modelled this presumption that more recent experience is more relevant for predicting the future by using geometrically declining weights in the vector autoregression that removes the serial correlation from the observed asset returns. For each element of the conditional variance–covariance structure, $\hat{\omega}_{ijt}$, the relevant expression is

$$\hat{\omega}_{ijt} = \frac{1 - \phi}{1 - \phi^{t-1}} \sum_{k=0}^{t-2} \phi^k \, \hat{\varepsilon}_{i,t-1-k} \, \hat{\varepsilon}_{j,t-1-k}$$

where $\hat{\varepsilon}_{i,t-1-k}$ is the inferred 'surprise' component of the return on the ith asset observed at time $t-1-k$, and ϕ is a 'discount' parameter such that ϕ close to unity means that information

98 *Benjamin M. Friedman*

remains relevant for a long time, while φ close to zero means that information quickly becomes useless.

The procedure used by Agell and Persson is simply the special case of this scheme in which φ is identically equal to one, so that all experience since the arbitrary beginning of the sample period is presumed to be equally relevant. Kuttner and I found that φ near unity enabled the inferred asset demands to deliver market equilibria corresponding to observed asset returns in the case of long-term debt, but not in the case of equity. For equity returns, the model most nearly matched observed experience with φ = 0.5 (a fairly small value in a quarterly context). In the end, of course, whether this or any similar procedure is useful depends in large part on how and why the conditional variance–covariance structure is changing.

9.4 Implications of Debt Management for Non-Financial Economic Activity

Given Agell and Persson's inferred conditional variance–covariance structure at each point in time, and the resulting inferred asset demands, a logical next step is to draw inferences about the effects of debt management actions on such aspects of economic activity as output, investment, employment, and price inflation. Agell and Persson are reluctant to do so for several reasons. They are concerned about holding fixed various responses of borrowers and lenders, various asset supplies, and various asset prices. They are concerned about the problem of Ricardian equivalence. They are concerned about the Lucas critique. And they report that the quantitative estimates of asset substitutabilities that they find are too small to admit major macroeconomic effects of pure asset exchanges anyway.

I have carried out an analysis along similar lines by imbedding something analogous to Agell and Persson's asset demand system within the MIT–Penn–SSRC (MPS) quarterly econometric model of the USA. The results of this exercise (see Part II below) are largely in line with what one might think, a priori, that debt management operations do. Increasing the outstanding supply of government short-term debt, and simultaneously reducing the supply of long-term debt, lowers the excess returns on both long-

Does Debt Management Matter? 99

term debt and equity. These asset return differences, in turn, affect real economic activity in two ways. First, the lower cost of capital, in the model's Jorgenson-style investment functions, stimulates both business capital spending and residential construction. Second, with a higher level of economic activity from the additional investment, corporations earn more profits. These higher profits get discounted into stock market prices—and, with a lower required return to equity, at a higher price–earnings ratio. With substantially higher stock prices, arising both from the higher price–earnings ratio and from greater profits, the model's life-cycle consumption function results in more consumption to reinforce the additional investment. In the end, therefore, an asset demand system like that of Agell and Persson leads to the result that substituting short- for long-term debt is expansionary, not only for investment (through cost-of-capital effects) but also for consumption (through life-cycle effects).

In their paper, Agell and Persson write carefully about the problems raised for this sort of analysis by both the Lucas critique and the Ricardian equivalence proposition. As they acknowledge, however, the purely financial modelling that they carry out in their paper is already subject to the Lucas critique. Moreover, the proposition that the public appropriately compensate for changes in the maturity structure of the outstanding government debt is even less credible than the proposition that the public compensate for changes in the outstanding total quantity—about which there is already a large and growing literature, at both theoretical and empirical levels.[5] With Ricardian equivalence dubious at best, and the analysis they have done already having crossed the bridge of rendering itself subject to the Lucas critique, there is little reason not to go ahead and examine the macroeconomic effects of potential debt management actions. After all, the possibility of such macro effects is what is most interesting about the subject of debt management policy in the first place.

Agell and Persson discuss the possibility of such effects but, in the end, dismiss them on the ground that their estimated derivatives showing the impact of pure asset exchanges on the

---

[5] See e.g. Bernheim's (1987) summary of relevant theoretical and empirical work.

100   *Benjamin M. Friedman*

market-clearing structure of *ex ante* excess returns are too small (in absolute value) to lead to material non-financial impacts.[6] Wholly apart from the distinction that they point out between new issues and the outstanding stock of securities, even the quantitative estimates that they report are not so small as to be dismissed totally. They report, for example, that 'increasing the share of long-term bonds by one percentage point will, using the conditional covariances for 1988(2), raise the expected bond yield by *merely* 0.0476 percentage points' (emphasis added). While a 5-basis-point impact for every 1 per cent change in the share of bonds is smaller than the estimates that I have reported in my work, in the context of recent actual changes in the amount and composition of the US government's outstanding debt, they are not trivial either (see Friedman 1985b).

One key determinant of the magnitude of such derivatives, as Frankel (1985) has emphasized, is the relevant set of assets construed to constitute total wealth. A matter that frequently arises in this regard is whether to include only financial assets (as in Agell and Persson's paper), or also physical assets (as in Frankel's). I have attempted to incorporate owner-occupied houses and consumer durables in an analysis along just these lines (Friedman 1985a). The results were nonsensical in several important respects, suggesting that portfolio considerations of risk and return are not sufficient to model the demand for physical assets that bear substantial non-pecuniary service returns. The principal implication here is that a narrower rather than a broader construction of aggregate wealth is warranted, and hence that derivatives like those reported here are likely to be larger rather than smaller for any given size of debt management action.

By contrast, an issue that cuts in the other direction is the relevance of foreign assets under conditions of international capital mobility. Indeed, given that the intended ultimate policy relevance of Agell and Persson's paper is presumably to Sweden (even though they carry out the analysis using US data), it is surprising that they did not consider this question. I have found in my work on the USA that the short-term debt instruments of Canada, West Germany, Japan, and the UK are, in each case,

---

[6] In this regard, their results parallel those of Frankel (1985).

Does Debt Management Matter? 101

close substitutes for US short-term debt instruments even on an uncovered basis (see Friedman 1987). I suspect that a corresponding analysis for Sweden would deliver similar or even stronger results.

9.5 A More Fundamental Issue in Modelling Asset Returns

In modelling the behaviour of investors forming expectations of asset returns, Agell and Persson implicitly follow the standard assumption that investors perceive the stochastic distribution of these returns as if, once per period, there is a drawing from a (joint) normal distribution characterized by some mean and some variance–covariance structure. Especially for equity returns, however, there is substantial evidence of leptokurtosis. Fama (1965) prominently called attention to the 'fat tails' problem (although it was known much earlier), and following his work there has been some tradition of using non-normal distributions, including in particular the stable Paretian distribution, for this purpose.

In recent work I have pursued the possibility that some of the problems encountered in many researchers' earlier work on this subject, and perhaps some that Agell and Persson have encountered in their paper as well, come from just this assumption— that returns behave as if they were drawn, once per period, from a single stable distribution (see Friedman and Laibson 1989). Following earlier work at the theoretical level, and more recently some work at the empirical level as well, David Laibson and I have approached this problem by using a combination of normal and Poisson distributions.[7] In brief, our central assumption is that the return realized in each period is the sum of a single drawing from a normal distribution, as usual, plus n drawings from a distinct normal distribution, where n differs from period to period and is, in each period, the outcome of a Poisson process. If this Poisson process has small enough λ, then, instead of n

---

[7] Early theoretical contributions along these lines were Merton (1971, 1976) and Cox and Ross (1976). Examples of recent empirical applications include Akgiray and Booth (1988), Ahn and Thompson (1988), Ball and Torous (1983), Clark (1973), Feinstone (1985), Jarrow and Rosenfeld (1984), Oldfield *et al.* (1977), and Tucker and Pond (1988).

102   *Benjamin M. Friedman*

being greater than one, it is typically zero; instead of many drawings from the second normal distribution in each period, there is a drawing from it only once every so many periods. In fact, our work on quarterly US equity returns indicates that, over the 28-year span examined by Agell and Persson, there have been only four drawings from this second normal distribution.

How does all this matter for what Agell and Persson do in their paper? As I have emphasized, their work here hinges crucially on the serial correlation properties of the returns that they analyse. If, however, the variance in these return data is dominated by a few large 'bangs' (an example would be the stock market crash in October 1987), and if these 'bangs' follow a Poisson process, then rational investors understand that, just because there was such a 'bang' last period, there is no reason to believe that one will occur next period. At some randomly determined time in the future, another one is likely to occur; but the probability of there being such a 'bang' in 1988 was no greater (or smaller) just because there was one in 1987.

If asset returns in fact follow such a mixed Poisson process, then any methodology like Agell and Persson's (or mine in previous work), which forces all return realizations into one normal distribution and then takes out the serial correlation, will distort the resulting conditional risk perceptions in a variety of ways. If there is a Poisson element to asset returns, forcing all realizations into one serial correlation structure is too restrictive, and doing so can lead to implausible results. What is perhaps most striking about the behaviour of actual investors, whenever one has the opportunity to observe it, is the richness (which is not the same as rationality) of the thought processes at work. By comparison, economists' formal structures can, at times, seem a poor skeleton.

Agell and Persson have done much to reawaken interest in debt management policy, and they have usefully added to the growing literature of empirical research on asset demand behaviour and expectations formation. To recognize that there is much yet to be done is not to belie their substantial contribution.

References for Part I

Agell, J. and Persson, M. (1987), 'Samhällsekonomiska effekter av statsskuldspolitiken' (Economic Effects of Government Debt Management), Appendix no. 7 to the Swedish Government's *1987 Medium Term Survey of the Swedish Economy* (LU 87). Stockholm.

Ahn, C. M. and Thompson, H. E. (1988), 'Jump–Diffusion Processes and the Term Structure of Interest Rates', *Journal of Finance*, 43: 155–74.

Akgiray, V. and Booth, G. G. (1988), 'Mixed Diffusion–Jump Process Modeling of Exchange Rates', *Review of Economics and Statistics*, 70: 631–7.

Auerbach, A. J. and King, M. A. (1983), 'Taxation, Portfolio Choice, and Debt–Equity Ratios: A General Equilibrium Model', *Quarterly Journal of Economics*, 98(4): 587–609.

Backus, D., Brainard, W., Smith, G. and Tobin, J. (1980), 'A Model of US Financial and Nonfinancial Economic Behavior', *Journal of Money, Credit and Banking*, 12: 239–93.

——and Purvis, D. (1980), 'An Integrated Model of Household Flow-of-Funds Allocations', *Journal of Money, Credit and Banking*, 12: 400–21.

Ball, C. A., and Torous, W. N. (1983), 'A Simplified Jump Process for Common Stock Returns', *Journal of Financial and Quantitative Analysis*, 18: 53–65.

Barro, R. J. (1974), 'Are Government Bonds Net Wealth?' *Journal of Political Economy*, 82: 1095–118.

Bergstrom, A. R. (1984), 'Continuous-Time Stochastic Models and Issues of Aggregation Over Time', in Z. Griliches and M. D. Intriligator (eds.), *Handbook of Econometrics*, ii. Amsterdam: North-Holland.

Berndt, E. R. (1991), 'Causality and Simultaneity Between Advertising and Sales', ch. 8 in *The Practice of Econometrics Classic and Contemporary*. Reading, Mass.: Addison-Wesley.

Bernheim, B. D. (1987), 'Ricardian Equivalence: An Evaluation of Theory and Evidence', in S. Fischer (ed.), *NBER Macroeconomics Annual*. Cambridge, Mass.: MIT Press.

Black, F. (1976), 'The Pricing of Commodity Contracts', *Journal of Financial Economics*, 3: 167–79.

104 *References for Part I*

—— and Scholes, M. (1973), 'The Pricing of Options and Corporate Liabilities', *Journal of Political Economy*, 81: 637–54.

Blanchard, O. and Plantes, M. (1977), 'A Note on Gross Substitutability of Financial Assets', *Econometrica*, 45: 769–71.

Bodie, Z. (1986), 'Investment Strategy in an Inflationary Environment', in B. M. Friedman (ed.), *The Changing Roles of Debt and Equity in Financing US Capital Formation*. Chicago: University of Chicago Press.

—— Kane, A. and McDonald, R. (1983), 'Why Are Real Interest Rates So High?' NBER Working Paper no. 1141, June.

Bollerslev, T., Engle, R. F. and Wooldridge, J. M. (1988), 'A Capital Asset Pricing Model with Time-Varying Covariances', *Journal of Political Economy*, 96: 116–31.

Box, G. E. P. and Jenkins, G. M. (1970), *Time Series Analysis*. San Francisco: Holden-Day.

Brainard, W. and Tobin, J. (1968), 'Pitfalls in Financial Model Building', *American Economic Review*, 58: 99–122.

Brownlee, O. H. and Scott, I. O. (1963), 'Utility, Liquidity and Debt Management', *Econometrica*, 31(3): 349–62.

Chamley, C. and Polemarchakis, H. (1984), 'Assets, General Equilibrium and the Neutrality of Money', *Review of Economic Studies*, 51(1): 129–38.

Chan, L. K. C. (1983), 'Uncertainty and the Neutrality of Government Financing Policy', *Journal of Monetary Economics*, 11: 351–72.

Chouraqui, J.-C., Jones, B. and Montador, R. B. (1986), 'Public Debt in a Medium-Term Perspective', *OECD Economic Studies* no. 7.

Clark, P. K. (1973), 'A Subordinated Stochastic Process Model with Finite Variance for Speculative Prices', *Econometrica*, 41: 135–55.

Cox, J. C., Ingersoll, J. E. Jun. and Ross, S. A. (1985a), 'An Intertemporal General Equilibrium Model of Asset Prices', *Econometrica*, 53(2): 363–84.

—— —— —— (1985b), 'A Theory of the Term Structure of Interest Rates', *Econometrica*, 53(2): 385–407.

—— and Ross, S. A. (1976), 'The Valuation of Options for Alternative Stochastic Processes', *Journal of Financial Economics*, 3: 145–66.

De Long, J. B., Shleifer, A., Summers, L. H., and Waldmann, R. J. (1990), 'Noise Trader Risk in Financial Markets', *Journal of Political Economy*, 98: 703–38.

Engel, C., Frankel, J., Froot, K. and Rodriguez, A. (1989), 'Conditional Mean–Variance Efficiency of the US Stock Market'. NBER Working Paper no. 2890, March.

—— and Rodrigues, A. (1989), 'Tests of International CAPM with Time-Varying Covariances', *Journal of Applied Econometrics*, 4: 119–38.

Does Debt Management Matter? 105

Engle, R. F. (1982), 'Autoregressive Conditional Heteroscedasticity with Estimates of the Variance of United Kingdom Inflation', *Econometrica*, 50: 987–1007.

Fair, R. and Malkiel, B. (1971), 'The Determination of Yield Differentials between Debt Instruments of the Same Maturity', *Journal of Money, Credit and Banking*, 3: 733–49.

Fama, E. F. (1965), 'The Behavior of Stock Market Prices', *Journal of Business*, 38: 34–105.

Feinstone, L. J. (1985), 'Minute by Minute: Efficiency, Normality, and Randomness in Intradaily Asset Prices'. Mimeo, University of Rochester.

Fischer, S. (1983), 'Investing for the Short and the Long Term', in Z. Bodie and J. B. Shoven (eds.), *Financial Aspects of the United States Pension System*. Chicago: University of Chicago Press.

——(1986), *Indexing, Inflation, and Economic Policy*. Cambridge, Mass.: MIT Press.

—— and Pennacchi, G. (1985), 'Serial Correlation of Asset Returns and Optimal Portfolios for the Long and Short Term'. Mimeo, National Bureau of Economic Research.

Frankel, J. (1982), 'In Search of the Exchange Risk Premium: A Six-Currency Test of Mean–Variance Efficiency', *Journal of International Money and Finance*, 2: 255–74.

——(1985), 'Portfolio Crowding-Out Empirically Estimated', *Quarterly Journal of Economics*, 100: 1041–66.

——(1991), 'Quantifying International Capital Mobility in the 1980s', in D. Bernheim and J. Shoven (eds.), *National Saving and Economic Performance*. Chicago: University of Chicago Press.

Friedman, B. M. (1977), 'Financial Flow Variables and the Short-Run Determination of Long-Term Interest Rates', *Journal of Political Economy*, 75: 661–89.

——(1978), 'Crowding Out or Crowding In? Economic Consequences of Financing Government Deficits', *Brookings Papers on Economic Activity*, 3: 593–641.

——(1985a), 'Portfolio Choice and the Debt-to-Income Relationship', *American Economic Review*, 75: 338–43.

——(1985b), 'Crowding Out or Crowding In? Evidence on Debt–Equity Substitutability'. Mimeo, National Bureau of Economic Research.

——(1986), 'Implications of Government Deficits for Interest Rates, Equity Returns, and Corporate Financing', in B. M. Friedman (ed.), *Financing Corporate Capital Formation*. Chicago: University of Chicago Press.

——(1987), 'The Substitutability of International Assets'. Mimeo, Harvard University.

106   References for Part I

—— and Kuttner, Kenneth N. (1992), 'Time-Varying Risk Perceptions and the Pricing of Risky Assets'. *Oxford Economic Papers*, forthcoming.

—— and Laibson, D. I. (1989), 'Economic Implications of Extraordinary Movements in Stock Prices', *Brookings Papers on Economic Activity*, 2: 137–89.

—— and Roley, V. V. (1987), 'Aspects of Investor Behavior Under Risk', In G. R. Feiwel (ed.), *Arrow and the Ascent of Modern Economic Theory*. London: Macmillan.

Friend, I. and Blume, M. E. (1975), 'The Demand for Risky Assets'. *American Economic Review*, 65: 900–22.

Froot, K. (1989), 'New Hope for the Expectations Hypothesis of the Term Structure of Interest Rates', *Journal of Finance*, 44: 283–305.

—— and Frankel, J. (1989), 'Forward Discount Bias: Is It an Exchange Risk Premium?', *Quarterly Journal of Economics*, 104: 139–61.

Grossman, S. and Shiller, R. (1981), 'The Determinants of the Variability of Stock Market Prices', *American Economic Review*, 71: 222–7.

—— Melino, A., and Shiller, R. J. (1985), 'Estimating the Continuous Time Consumption Based Asset Pricing Model'. NBER Working Paper no. 1643.

IMF (various years), *International Financial Statistics*. Washington: IMF.

Jarrow, R. A. and Rosenfeld, E. R. (1984), 'Jump Risks and the Intertemporal Capital Asset Pricing Model', *Journal of Business*, 57: 337–51.

Jorgenson, D. W. and Yun, K.-Y. (1986), 'Tax Policy and Capital Allocation', *Scandinavian Journal of Economics*, 88(2): 355–77.

King, M. A. and Leape, J. (1984), 'Wealth and Portfolio Composition: Theory and Evidence', *Economic and Social Research Council Programme in Taxation, Incentives and the Distribution of Income*, 68. London: ESRC.

Lucas, R. E. Jun. (1976), 'Econometric Policy Evaluation: A Critique', in K. Brunner and A. Meltzer (eds.), 'The Phillips Curve and Labor Markets', Supplement to *Journal of Monetary Economics*.

—— and Stokey, N. L. (1983), 'Optimal Fiscal and Monetary Policy in an Economy Without Capital', *Journal of Monetary Economics*, 12: 55–93.

Markowitz, H. M. (1959), *Portfolio Selection*. New York: John Wiley.

Masson, P. (1978), 'Structural Models of the Demand of Bonds and the Term Structure of Interest Rates', *Economica*, 45: 363–77.

Mayers, D. (1972), 'Non-Marketable Assets and the Capital Market Equilibrium under Uncertainty', in M. C. Jensen (ed.), *Studies in the Theory of Capital Markets*. New York: Praeger.

Does Debt Management Matter? 107

McDonald, R. L. (1983), 'Government Debt and Private Leverage: An Extension of the Miller Theorem', *Journal of Public Economics*, 22(3): 303–25.

Mehra, R. and Prescott, E. C. (1985), 'The Equity Premium: A Puzzle', *Journal of Monetary Economics*, 15: 145–61.

Merton, R. C. (1971), 'Optimum Consumption and Portfolio Rules in a Continuous-Time Model', *Journal of Economic Theory*, 3: 373–413.

—— (1976), 'Option Pricing When Underlying Stock Returns Are Discontinuous', *Journal of Financial Economics*, 3: 125–44.

—— (1982), 'On the Microeconomic Theory of Investment Under Uncertainty', in K. J. Arrow and M. D. Intriligator (eds.), *Handbook of Mathematical Economics*, ii. Amsterdam: North-Holland.

Modigliani, F. and Miller, M. H. (1958), 'The Cost of Capital, Corporation Finance, and the Theory of Investment', *American Economic Review*, 48: 261–97.

—— and R. Sutch (1966), 'Innovations in Interest Rate Policy', *American Economic Review*, 56: 178–97.

—— —— (1967), 'Debt Management and the Term Structure of Interest Rates: An Empirical Analysis of Recent Experience', *Journal of Political Economy*, 75: 569–89.

Musgrave, R. A. (1959), *The Theory of Public Finance*, New York: McGraw-Hill.

Oldfield, G. S. Jun., Rogalski, R. J., and Jarrow, R. A. (1977), 'An Autoregressive Jump Process for Common Stock Returns', *Journal of Financial Economics*, 5: 389–418.

Okun, A. (1963), 'Monetary Policy, Debt Management, and Interest Rates: A Quantitative Appraisal', in *Fiscal and Debt Management Policies*. Englewood Cliffs, NJ: Prentice-Hall.

Peled, D. (1985), 'Stochastic Inflation and Government Provision of Indexed Bonds', *Journal of Monetary Economics*, 15(3): 291–308.

Persson, M., Persson, T. and Svensson, L. E. O. (1987), 'Time Consistency of Fiscal and Monetary Policy', *Econometrica*, 55(6): 1419–32.

Phelps, E. S. (1973), 'Inflation in the Theory of Public Finance', *Swedish Journal of Economics*, 75: 67–83.

Pindyck, R. (1984), 'Risk, Inflation and the Stock Market', *American Economic Review*, 74: 335–51.

Poterba, J. M., and Summers, L. H. (1987), 'Mean Reversion in Stock Prices: Evidence and Implications', *Journal of Financial Economics*, 22: 27–59.

Roley, V. V. (1979), 'A Theory of Federal Debt Management', *American Economic Review*, 69: 915–26.

—— (1982), 'The Effect of Debt Management Policy on Corporate Bond

References for Part I

and Equity Yields', *Quarterly Journal of Economics*, 97: 645–68.

Rolph, E. R. (1957), 'Principles of Debt Management', *American Economic Review*, 47: 302–20.

Smith, G. and Brainard, W. (1976), 'The Value of A Priori Information in Estimating a Financial Model', *Journal of Finance*, 31: 1299–1322.

Stiglitz, J. E. (1983), 'On the Relevance or Irrelevance of Public Financial Policy: Indexation and Optimal Monetary Policy'. NBER Working Paper no. 1106.

Tobin, J. (1958), 'Liquidity Preference as Behavior Towards Risk', *Review of Economic Studies*, 25: 65–86; reprinted in J. Tobin, *Essays in Economics*, i. Amsterdam: North-Holland, 1971.

—— (1963), 'An Essay on the Principles of Debt Management', in *Fiscal and Debt Management Policies*. Englewood Cliffs, NJ: Prentice-Hall; reprinted in J. Tobin, *Essays in Economics*, i. Amsterdam: North-Holland, 1971.

—— (1969), 'A General Equilibrium Approach to Monetary Theory', *Journal of Money, Credit and Banking*, 1: 15–29; reprinted in J. Tobin, *Essays in Economics*, i. Amsterdam: North-Holland, 1971.

Tucker, A. L. and Pond, L. (1988), 'The Probability Distribution of Foreign Exchange Price Changes: Tests of Candidate Processes', *Review of Economics and Statistics*, 70: 638–47.

Wallace, N. (1981), 'A Modigliani–Miller Theorem for Open-Market Operations', *American Economic Review*, 71(3): 267–74.

Werin, L. (1990), 'An Applied General Equilibrium Model of the Asset Markets in Sweden'. In L. Bergman, D. Jorgenson and E. Zalai (eds.), *General Equilibrium Modeling and Economic Policy Analysis*. Oxford: Basil Blackwell.

Yarrow, G. (1986), 'Privatization in Theory and Practice', *Economic Policy*, no. 2: 324–64.

PART II

Debt Management Policy, Interest Rates, and Economic Activity

10

Introduction

At the end of the Second World War, the US Treasury's outstanding debt was mostly long-term: $54 billion of bonds maturing in ten years or more, $66 billion of notes and bonds maturing in one to ten years, and only $47 billion of bills and other securities due in less than one year, amounting to an overall mean maturity of 116 months. In financing new deficits and refinancing maturing issues during the first three decades of the postwar period, the Treasury usually relied on shorter-term borrowing, thereby substantially reducing the outstanding debt's average term to maturity. At times this pattern of debt management resulted from the statutory limitations (now relaxed) on issuing bonds bearing coupons greater than $4\frac{1}{4}$ per cent, but at other times it also reflected discretionary Treasury policy. Since 1975, however, the Treasury's management of its debt has taken the opposite tack (see Table 10.1).[1]

The mean maturity of the Treasury's outstanding debt reached 28 months, its postwar minimum, in January 1976. Since then reliance on regular medium-term note issues and 20- and 30-year bonds has extended the mean maturity to 71 months as of year-end 1990. Hence the result of the Treasury's debt management

I am grateful to Arturo Estrella, Orlin Grabbe, David Johnson, and Richard Mattione for research assistance; to them as well as James Duesenberry, John Lintner, Franco Modigliani, James Pesando, Vance Roley, James Tobin, and Edwin Yeo for helpful discussions and comments on earlier drafts; and to the National Science Foundation, the Alfred P. Sloan Foundation, and the Harvard Program for Financial Research for research support. I am especially grateful to Vance Roley for generous advice on the adaptation of his model for use in this paper.

[1] The distribution data in Table 10.1 slightly *under*state the shift to shorter maturities because of the switch from a first-call classification for 1945–55 to a final-maturity classification for 1960–90. (The first-call breakdown for 1960, corresponding to that shown in the table, is 43.0%, 39.7%, 9.6%, 2.4%, 5.2%.) The mean maturity computation is based on final maturity and is consistent throughout.

TABLE 10.1 *Maturity structure of marketable interest-bearing US Treasury debt held by private investors, 1945–1990*

Year-end	Total ($bn.)	Within 1 yr. (%)	1–5 yrs. (%)	5–10 yrs. (%)	10–20 yrs. (%)	20 yrs. and over (%)	Mean maturity (mos.)
1945	167.5	28.1	20.5	19.1	19.2	13.1	116
1950	126.3	33.2	25.3	12.7	28.9	0.0	100
1955	134.2	34.0	29.8	25.7	7.5	3.0	71
1960	153.5	38.2	37.5	10.4	7.5	6.4	58
1965	160.4	41.9	27.0	18.8	3.9	8.4	63
1970	168.5	49.9	33.9	7.5	3.9	4.8	41
1975	255.9	58.7	29.2	6.5	3.3	2.3	29
1980	492.3	48.7	32.4	8.4	5.5	5.0	45
1985	1237.3	39.6	34.2	13.2	5.3	7.6	60
1990	1925.4	34.6	34.3	14.0	4.5	12.5	71

Notes:
Callable securities classified by date of first call for 1945–55, by final maturity for 1960–90.
Data for end of calendar year.

Source: US Department of the Treasury (*Treasury Bulletin*, Office of the Secretary of the Treasury, Office of Government Financing)

Debt Management Policy and Interest Rates   113

policy since 1976 has been to lengthen the debt just as rapidly
as the average policy prevailing during the previous 30 years
shortened it. The only significant postwar episode of debt lengthening before 1976 was during the early 1960s, when the mean
maturity rose from 53 months in January 1960 to 69 months in
June 1965. By contrast, the more recent maturity lengthening
has lasted three times as long, and has been much more
extensive.

Do changes in Treasury debt management affect the financial
markets or the non-financial economy? Although many economists assumed so early in the postwar period, a series of empirical
investigations beginning in the 1960s provided either weak evidence for such effects or none at all. More recently, however,
researchers using explicit demand-and-supply models of interest
rate determination have provided partial equilibrium evidence
of sufficiently imperfect asset substitutabilities in private-sector
portfolios to suggest substantial effects associated with major
debt management actions. Still, the effects of such actions
on economic activity in a general equilibrium context remain
unexplored.

The object of this paper is to provide a quantitative assessment
of the economic effects of debt management policy based on an
explicit demand-and-supply model of interest rate determination.
The principal research tool employed here for this purpose is a
hybrid model combining the familiar MIT–Penn–SSRC (henceforth MPS) econometric model of the USA, a structural model
developed in Friedman (1977, 1979) representing the determination of interest rates and financing volume in the US corporate
bond market, and a structural model developed in Roley (1980,
1982) representing the determination of interest rates in four
separate maturity sub-markets of the US government securities
market. The basis of the two interest rate models is the requirement that, in each asset market, the amount of securities
demanded by investors must equal the amount supplied by
borrowers—including either private-sector borrowers or the
government. Hence changes in the pattern of government debt
management can directly affect the market-clearing structure of
yields. The MPS model (minus its term-structure equation, which
is redundant in the presence of the structural interest rate
models) in turn develops the implications of these yield move-

114   *Benjamin M. Friedman*

ments for other aspects of financial as well as non-financial economic activity. Moreover, because the combined model is fully simultaneous, the general equilibrium solution that it determines allows for a rich set of feedbacks in both directions between financial and non-financial aspects of economic behaviour.

Chapter 11 reviews the underlying theory relating the structure of asset yields generally, and the volumes of inside assets supplied and demanded, to the relative supplies of outside assets. Chapter 12 describes the combined MPS and demand–supply interest rate model used for the empirical analysis. Chapter 13 reports simulation experiments assessing the effects of two different debt management actions—one a sustained change in the pattern of new financing, and the other a larger change in the maturity structure effected within one year. To anticipate, the results of these experiments indicate that such debt management actions, in plausible magnitudes, would have significant effects not only on the structure of asset yields and prices but also on both the level and the composition of non-financial economic activity. Chapter 14 briefly summarizes the paper's principal conclusions.

11

Debt Management, Interest Rates, and Asset Prices

When asset markets are in equilibrium, the market-clearing structure of returns depends in a straightforward way on the quantities of outside assets supplied. If investors' preferences exhibit constant relative risk aversion, and if their assessments of the returns on the available assets are normally (or joint normally) distributed, in the absence of transaction costs their optimal single-period portfolio allocation will be (approximately) of the form

$$\mathbf{A}_t^D = W_t(B_t \mathbf{r}_t^e + \boldsymbol{\pi}_t), \tag{1}$$

where $\mathbf{A}^D$ is a vector of asset demands satisfying $\mathbf{A}^{D\prime}\mathbf{1} = W$, W is total portfolio wealth, $\mathbf{r}^e$ is a vector of means of the joint asset return distribution corresponding to $\mathbf{A}^D$, B and $\boldsymbol{\pi}$ are respectively a matrix and a vector of coefficients determined by the coefficient of relative risk aversion and the variance–covariance matrix of the asset return distribution, and t denotes the tth time period.[1] Because $\mathbf{A}^D$ is proportional to W and linear in $\mathbf{r}^e$, (1) is both the optimal allocation for a single investor when W is that one investor's wealth, and also the economy-wide optimal allocation

---

[1] The specific form of (1), if all assets are risky, is $B = (1/\rho)[\Omega^{-1} - (\mathbf{1}'\Omega^{-1}\mathbf{1})^{-1}\Omega^{-1}\mathbf{1}\mathbf{1}'\Omega^{-1}]$ and $\boldsymbol{\pi} = (\mathbf{1}'\Omega^{-1}\mathbf{1})^{-1}\Omega^{-1}\mathbf{1}$, where ρ is the coefficient of relative risk aversion and Ω is the variance–covariance matrix. Here B is singular, so that the asset demand system will be capable of determining all relative yields and all but one absolute yield. Alternatively, in the presence of a risk-free (certain return) asset the full Ω matrix is singular, so that it is necessary to partition the set of demands; the resulting asset demand system, in which $\mathbf{A}^D$, $\mathbf{r}^e$, and Ω refer to the risky assets only, is then just $\mathbf{A}^D = W(B\mathbf{r}^e)$ where $B = (1/\rho)\Omega^{-1}$, and the optimal portfolio demand for the risk-free asset is simply $(W - \mathbf{A}^{D\prime}\mathbf{1})$. See Friedman and Roley (1987) for the result that constant relative risk aversion and joint normal asset return assessments imply (approximate) asset demand functions that are homogeneous in wealth and linear in expected returns. (The combination of constant relative risk aversion and normal distributions is only an approximation, of course, in that the underlying utility function is undefined for negative wealth values.)

116   *Benjamin M. Friedman*

when W is aggregate wealth and all investors have identical preferences and assessments. The economy-wide optimal allocation is still of the form (1) even if investors exhibit heterogeneous preferences or hold diverse assessments, and the aggregate B and π in this case are combinations of the B and π appropriate for the underlying individuals, weighted by their respective individual wealth totals.[2]

The partial equilibrium of the asset markets is equivalent to the market-clearing condition

$$\mathbf{A}_t^D = \mathbf{A}_t^S \tag{2}$$

where $\mathbf{A}^S$ is a vector of given net asset supplies, including non-zero values for all existing outside assets and zero values for all inside assets. Substitution from (1) then yields the determination of expected asset returns according to[3]

$$\mathbf{r}_t^e = B_t^{-1} \left(\frac{1}{W} \mathbf{A}_t^S - \pi_t \right). \tag{3}$$

For given wealth, and given preferences and variance–covariance assessments (so that B and π are fixed), variations in outside asset supplies clearly affect the market-clearing structure of returns. The role of government debt management policy in this context is, in the first instance, to achieve just such effects.

For example, consider the case of a model with no inside assets and only four outside assets: money (M), short-term government debt (S), long-term government debt (L), and capital (K). Dividing both sides of (2) by wealth (defined here as $W = M + S + L + K$) and applying the relevant symmetry and balance sheet constraints facilitate representing the structure of the asset markets in terms of the six off-diagonal elements of B indicating the four assets' respective pairwise substitutabilities:[4]

---

[2] See Lintner (1969) and Friedman (1980) for explicit treatments of the case of heterogeneous investors.

[3] Because the full B matrix is singular (see again fn. 1), the expression in (3) has dimension reduced by one and therefore represents the determination of relative yields against an arbitrary fixed benchmark in the case of all risky assets or against the certain yield in the presence of a risk-free asset. An isomorphic interpretation of (3) is that relative asset returns depend on shares in the market portfolio relative to shares in the minimum-variance portfolio.

[4] Including an additional term to represent the dependence of the demand for money (and hence at least one other asset) on income, in accordance with

$$
\begin{vmatrix} 1-\pi_s-\pi_l-\pi_k \\ \pi_s \\ \pi_l \\ \pi_k \end{vmatrix}
+
\begin{vmatrix}
-b_{ms}-b_{ml}-b_{mk} & b_{ms} & b_{ml} & b_{mk} \\
b_{ms} & -b_{ms}-b_{sl}-b_{sk} & b_{sl} & b_{sk} \\
b_{ml} & b_{sl} & -b_{ml}-b_{sl}-b_{lk} & b_{lk} \\
b_{mk} & b_{sk} & b_{lk} & -b_{mk}-b_{sk}-b_{lk}
\end{vmatrix}
\begin{vmatrix} r_m \\ r_s \\ r_l \\ r_k \end{vmatrix}^e
=
\begin{vmatrix} M/W \\ S/W \\ L/W \\ K/W \end{vmatrix}. \quad (4)
$$

Debt management actions in this simple model consist of off-setting changes in the supplies of short- and long-term bonds which, at least to a first approximation, leave total wealth fixed—in other words, $dS = -dL$.[5]

What effect will such actions have on the resulting structure of asset returns? The form of the asset demand system in (4), which explicitly incorporates the balance sheet constraint, makes clear that only three of the four equations are independent, so that the system can determine only three of the four asset yields. With r_m fixed, for example, any three equations of the system can determine r_s, r_l, and r_k. Moreover, although the absolute levels of these three yields will depend on the π_i coefficients and on the fixed level of r_m, the marginal effect of any $dS = -dL$ on these three yields will depend only on the Jacobian B—which, as (4) shows, can be expressed completely in terms of the relevant asset substitutabilities. Solution of (4), for W, M, K, and r_m held fixed, yields these marginal effects as

---

standard transactions-inventory models, would not alter the analysis here; see Friedman (1978). See Roley (1983) and Friedman and Roley (1987) on the implications of a symmetric Jacobian.

[5] The condition $dS = -dL$ holds exactly in a timeless abstraction in which, with no previous history of assets outstanding, the government distributes one kind of bond or the other, or both. In a more realistic context, however, wealth does change because of valuation changes on the outstanding long-term bonds. For example, to anticipate the analysis that follows, suppose that the government issues short-term bonds and uses exactly the entire proceeds to buy back some of its outstanding long-term bonds. If that action reduces the yield on the long-term bonds still outstanding, the associated rise in the price of these bonds will raise total wealth. The reasoning is analogous to that argued below for the case of valuation of capital. See Roley (1979) for a theoretical discussion emphasizing changes in bond prices resulting from debt management actions.

118   *Benjamin M. Friedman*

$$\frac{dr_s}{dS} = \frac{b_{ml}(b_{mk} + b_{sk} + b_{lk}) + b_{mk}b_{lk}}{W\Delta}$$

$$\frac{dr_l}{dS} = -\frac{b_{ms}(b_{mk} + b_{sk} + b_{lk}) + b_{sk}b_{lk}}{W\Delta} \qquad (5)$$

$$\frac{dr_k}{dS} = \frac{b_{ml}b_{sk} - b_{ms}b_{lk}}{W\Delta},$$

where W is again total wealth, and Δ is the determinant of the subsystem of the matrix in (4) formed by eliminating the first column (since r_m is fixed) and any one row chosen arbitrarily.

Although the derivatives in (5) remain unsigned in the absence of any restrictions at all on B, the assumption that the four assets are gross substitutes—that is, that each off-diagonal b_{ij} coefficient is negative—renders the determinant unambiguously positive and therefore yields the intuitively plausible result,[6]

$$\frac{dr_s}{dS} > 0$$

$$\frac{dr_l}{dS} < 0 \qquad (6)$$

$$\frac{dr_k}{dS} \gtrless 0.$$

Hence a simultaneous sale of short-term bonds and purchase of long-term bonds unambiguously increases the yield on the former and reduces that on the latter, while still leaving ambiguous the effect on the asset with supply held constant. From (5), however, it is clear that a further intuitive assumption about the ordering of relative substitutabilities among asset pairs is sufficient to sign the effect on r_k also. If investors perceive an asset hierarchy (in terms of safety, liquidity, etc.) extending in order from money to short-term bonds to long-term bonds to capital, such that they regard adjacent assets in this ordering as better substitutes than more distant assets, then b_{ms} and b_{lk} will dominate b_{ml} and b_{sk}, and the

[6] See Blanchard and Plantes (1977) for a statement of necessary and sufficient conditions for gross substitutability. The covariance matrices reported in Friedman (1985a, b) suggest that, for broad asset categories like those under consideration here, at least the necessary conditions are met in practice.

Debt Management Policy and Interest Rates   119

effect of such a debt management action also unambiguously reduces the required yield on capital:

$$\frac{\mathrm{d}r_k}{\mathrm{d}S} < 0. \tag{6$'$}$$

Although the simple analysis in (1)–(6) sets forth the central idea behind the role of debt management policy in affecting asset returns, two generalizations of this analysis are important for the effects of debt management more broadly as investigated in Chapter 13 below.

First, although the general model in (1)–(3) in principle includes inside assets, the model as written says nothing about the determination of their respective quantities, and the special case considered in (4)–(6) excludes inside assets altogether. If all investors are identical, then of course the outstanding amount of each inside asset must be exactly zero on a gross basis as well as on the net basis represented by supply vector $\mathbf{A}^S$. One investor would not borrow from or lend to another if both shared identical preferences, assessments, and endowments. In a world in which preferences, assessments, and endowments may differ, however—and in which legal and other institutional restrictions may importantly constrain the behaviour of asset market participants—borrowing and lending among individuals and firms in fact constitute much of the financial markets' everyday activity.

An alternative form of (2) that explicitly recognizes this heterogeneity is just

$$\sum_h \mathbf{A}_{ht}^D = \mathbf{A}_t^S, \tag{7}$$

where h denotes the hth investor (which may be an individual or an institution), and for the jth asset the market-clearing condition is correspondingly

$$\sum_h \mathbf{A}_{hjt}^D = \mathbf{A}_{jt}^S. \tag{8}$$

The special characteristic of an outside asset is that $\mathbf{A}_j^S > 0$ (if the asset exists) and (unless there are short sales) $\mathbf{A}_{hj}^D \gtrless 0$ for all h, subject of course to (8). By contrast, the special characteristic of an inside asset is that $\mathbf{A}_j^S = 0$ and $\mathbf{A}_{hj}^D \gtrless 0$, again subject to (8).

While a fully aggregated asset market model can never identify

120 *Benjamin M. Friedman*

the gross quantity of an inside asset, a disaggregated model can do so if the disaggregation is such as to distinguish positive from negative A_{hj}^D. By far the most familiar such disaggregation in macroeconomic models is that between the banking system and the non-bank public, which permits identification of inside money; but other inside assets may be worth identifying as well. To the extent that gross quantities of inside assets matter in addition to the associated yields, therefore—for example, if the quantity of mortgage credit borrowed and lent affects homebuilding apart from the mortgage interest rate, or if the level of household indebtedness affects consumer spending apart from the interest rate on consumer credit, or if the volume and pattern of corporate financing affects business investment apart from the interest rates on bonds and bank loans—an advantage of an appropriately disaggregated model is its ability to identify and determine these quantities, thereby facilitating the representation of their effects on economic activity.

A second important shortcoming of the simplified analysis in (1)–(6) is that it excludes induced asset price effects.[7] Although it is possible to imagine situations in which both the quantity and the price of capital would remain fixed despite movements in the return on capital, a more plausible treatment would allow for a price response (say, in the short run) or a quantity response (say, in the long run), or both. Under the fixed-quantity variable-price conditions assumed by Tobin (1969), for example, the supply of capital in (4) is qK/W, where K is now the physical quantity of capital and q is its market price in relation to replacement cost, which varies with the associated return according to

$$q = f(r_k), \qquad f' < 0. \tag{9}$$

Because wealth now depends on the price of capital, which changes whenever r_k changes, the resulting asset market equilibrium is richer than that in (4). A fall in r_k as in (6') when $dS > 0$ increases the supply of capital at market prices. It therefore increases wealth, and therefore increases the sum of the demands for all assets including capital itself. Under the assumptions that give rise to the effects of a debt management action on r_s and r_l as in (6) and r_k as in (6'), the consequence of a flexible

---

[7] See again fn. 5.

Debt Management Policy and Interest Rates   121

price of capital is a reduction in the absolute magnitude of each yield effect without any change in sign. In particular, following linearization of (1), the effect of $dS = -dL$ on r_k is

$$\frac{dr_k}{dS} = \frac{b_{ml}b_{sk} - b_{ms}b_{lk}}{\Delta - Kf'[\pi_s(b_{sl}b_{lk} + b_{sk}b_{mk} + b_{sl}b_{sk} + b_{sk}b_{lk})}$$
$$+ \pi_l(b_{sl}b_{sk} + b_{lk}b_{ms} + b_{lk}b_{sl} + b_{sk}b_{lk})$$
$$+ (1 - \pi_k)(b_{ms}b_{ml} + b_{ms}b_{sl} + b_{ms}b_{lk} + b_{sl}b_{ml}$$
$$+ b_{sl}b_{lk} + b_{sk}b_{ml} + b_{sk}b_{sl} + b_{sk}b_{lk})]. \qquad (10)$$

This expression differs from the corresponding equation in (5) only by the second denominator term, which, like the determinant, is unambiguously positive if all assets are gross substitutes and if investors are wealth-diversifiers. Hence the effect on r_k is again negative, though smaller in absolute value, and from (9) the effect on q is simply

$$\frac{dq}{dS} = f'\frac{dr_k}{dS} > 0. \qquad (11)$$

Generalizing the model to allow for a flexible price of capital does not qualitatively change the effect of debt management on the asset market partial equilibrium, therefore, but, like the generalization to identify gross quantities of inside asset stocks, it may have important implications for associated effects beyond the asset markets. Models relating business investment to the ratio of market price to replacement cost, either instead of or in addition to capital and other asset yields *per se*, are well known. In addition, because ownership of equity claims to capital bulks large in household portfolios, and because the variation over time in equity prices is typically much greater than that in other asset prices, the flexible price of capital in fact accounts for most of the observed variation in household wealth. To the extent that household wealth in turn affects consumption, as in the standard life-cycle model, effects of debt management actions on asset prices again may affect economic activity.

Once debt management affects economic activity, however, the likely feedback effects on the asset markets render any simple partial equilibrium analysis no longer adequate. Changes in incomes and spending may affect asset demands and the resulting market-clearing yields directly (as in the presence of a transactions demand for money), or through changes in borrowing that typically accompany certain types of expenditures, or through changes in wealth arising from induced saving or dissaving. Hence a model of general equilibrium, incorporating these and other influences in both directions between the financial and non-financial markets, is necessary.

12

A Model of Interest Rates and Economic Activity

Most familiar empirical models of interest rate determination preclude analysis of debt management effects because they rule out such effects at the outset. After first determining some key short-term interest rate from the interaction of monetary policy and the demand for money, most current models then proceed to determine the yield on any other asset, like the interest rate on long-term bonds or the dividend price yield on equities, from a single reduced-form term-structure equation estimated directly with the yield in question as the dependent variable. Equations of this kind have become standard since the work of Modigliani and Sutch (1966) and Modigliani and Shiller (1973).

Because such term-structure equations in principle represent partial reduced forms of some (usually unspecified) structure that may resemble (1)–(3) above, there is no a priori reason why they cannot include one or more variables representing outside asset supplies. Indeed, in the mid-1960s several researchers attempted to isolate effects of the then recent 'Operation Twist' surrogate debt management policy in just this way. Although a few analyses showed some evidence of effects on the yield structure arising from changing asset supplies, most did not, and therefore concluded that Operation Twist had been a failure—not surprisingly, since, despite the Federal Reserve System's limited attempt to shorten the average maturity of the privately held government debt via open market purchases, offsetting Treasury financing led instead to a net lengthening (see again Table 10.1).[1] The finally estimated form of the 'preferred habitat' model of Modigliani and Sutch, for example, in fact included no 'preferred habitat' terms.

---

[1] The most comprehensive attempt to find evidence of such effects was that of Modigliani and Sutch (1967). An example of results exhibiting such effects is in Okun (1963), but Okun concluded that the effects were small. See Friedman (1977b) for evidence from a more recent sample.

124   *Benjamin M. Friedman*

Subsequent work on interest rates within the single-equation, directly estimated reduced-form framework has largely followed the same path. Through most of the postwar period, either there was too little variation (around trend) in the relative supplies of short- versus long-term outside bonds, or else investors regarded debts of different maturity as too closely substitutable for the standard term-structure equation to detect any asset supply effects. Analogous work based on more recent data has shown some evidence of such effects, but to date the extent of variation present in the data apparently still precludes drawing sharp conclusions from such imprecise methods, or from attempts to infer asset substitutabilities by estimating conditional variance–covariance structures.[2]

By contrast, demand-and-supply models of asset return determination constructed explicitly in the form of (1)–(3) above provide a way of extracting more information from the available data, in that the underlying market-clearing structure constrains the way in which asset quantity variables enter the analysis. In the most general terms, the demand-and-supply model facilitates using the theory of portfolio behaviour to restrict the model's implied equations for relative returns, while imposing on the researcher the discipline of acknowledging explicitly that any factor hypothesized to affect relative returns can do so only by affecting some investor's asset demands (or some borrower's asset supplies). The asset demand equations in models of the US corporate and government bond markets, constructed in this way by Friedman (1977a, 1979) and Roley (1980, 1982), respectively, indicate that investors (and private-sector borrowers) regard assets of different maturity as less than perfect substitutes. Moreover, partial equilibrium simulations of these models indicate that, because of this imperfect substitutability, changes in asset supplies have sizeable effects on the structure of asset returns.

For at least two reasons, however, it is useful to go beyond such partial equilibrium analysis. First, debt management effects within the asset markets are of only limited interest. What presumably matters for public policy is the effects such actions have on non-financial economic activity. Second, partial equi-

---

[2] See e.g. the contrasting findings reported by Friedman (1985b) and Frankel (1985).

Debt Management Policy and Interest Rates 125

librium analysis of the asset markets necessarily holds fixed all aspects of non-financial activity, including those aspects that debt management actions may affect. Allowing for the associated set of feedbacks requires instead an analysis of the resulting general equilibrium.

The model employed here for this purpose consists of an altered MPS model (1978 version, before the deregulation of deposit markets and the introduction of new monetary policy procedures), from which the familiar single term-structure equation determining the corporate bond yield has been removed and into which a demand–supply model of interest rate determination in the corporate and government bond markets has been substituted.

The corporate bond yield is by far the most important asset yield in the MPS model from the perspective of implications for non-financial economic behaviour. In the first instance, the corporate bond yield exerts a major influence on business fixed investment in the model through its role in determining the user cost of capital. It also exerts an analogous influence on residential investment at only one step removed; in this case the relevant user cost depends on the mortgage yield, which in turn follows from the corporate bond yield via a simple term-structure-like relationship. In addition, the corporate bond yield influences both durable and non-durable consumer spending. The motivation underlying the determination of expenditures on consumer durables is again analogous to that for business and residential investment, although in this case the model actually uses a simplified function relating these expenditures directly to the corporate bond yield. The primary determinant of non-durable consumption is households' wealth, which consists in large part of equities;[3] the model determines the market value of equities as the value of dividend payments, in turn determined by a function in which the corporate bond yield is one direct argument, among several, divided by the dividend price yield, which also follows from the corporate bond yield via another simple term-structure-

---

[3] What matters in this context is the contribution of equities not to the level but to the variation of households' wealth. Given the great volatility of equity prices in contrast with the fixed-price nature of deposits, the typical 30% share of equities in households' total wealth greatly understates the role of equities in accounting for the variation of household wealth over time.

126   *Benjamin M. Friedman*

like relationship. Finally, within the model's representation of the financial markets, the corporate bond yield is a direct argument of the functions determining numerous other yield and quantity variables which in turn exercise diverse influences on non-financial behaviour (including, most importantly, a credit availability effect in the mortgage market).

The corporate bond market model used here consists of eight equations representing the respective net purchases of corporate bonds by six categories of bond investors (life insurance companies, other insurance companies, private pension funds, state and local government retirement funds, mutual savings banks, and households) and net sales of corporate bonds by two categories of bond issuers (domestic non-financial business corporations and finance companies).[4] The model's ninth, and final, equation is a market-clearing equilibrium condition analogous to (2), which requires the algebraic sum of the net purchases and sales by all categories of bond investors and bond issuers to sum to zero, and hence permits the model to determine the corporate bond yield as in (3).[5]

The government bond market model used here has four parts, corresponding to sub-markets for four separate maturity classes of US Treasury securities. These maturity classes are defined in terms of four distinct ranges (within 1 year, 2–4 years, 6–8 years, and over 12 years), with securities in the three remaining indeterminate areas allocated to the respective preceding and succeeding ranges according to a weighting scheme designed to avoid anomalous effects that would otherwise occur when large individual debt issues cross arbitrary classification boundaries. The model for each maturity sub-market consists of a set of demand equations representing the respective net purchases of

[4] See Friedman (1977a, 1979) for details of the estimation. The model takes as exogenous the net bond purchases and sales of all investors and issuers other than those noted above. The explicitly modelled investors and issuers accounted for about 95% and 90%, respectively, of all corporate bonds issued in the USA during the sample period used to estimate the model.

[5] The particular bond rate used in this model is the observed new-issue yield on long-term utility bonds rated Aa by Moody's Investor Service, Inc. An additional equation then determines the Aaa seasoned corporate yield, the bond rate used in the MPS model, as a simple direct function of the Aa new-issue yield and the longest-term government yield. Eliminating the Aaa seasoned yield altogether and using the Aa new-issue yield in its place would require re-estimating each MPS model equation in which the corporate bond rate appears.

Debt Management Policy and Interest Rates   127

Treasury securities in that maturity class by either nine or ten categories of investors (the six listed above as corporate bond investors plus commercial banks, savings and loan associations, state and local government general funds, and, for the two shorter maturity ranges only, domestic non-financial business corporations).[6] The supply of securities in each maturity sub-market is exogenous to the model, and in each case a market-clearing equilibrium condition analogous to (2) determines the associated yield as in (3).[7]

The specification of each investor group's respective demand for either corporate bonds or any maturity class of government securities combines the asset demand system (1) for given wealth, generalized to allow for influences on desired portfolio allocations arising from factors other than expected returns (e.g. expected price inflation), with an optimal marginal adjustment model that represents in a tractable way the effect of differential transaction costs that render the allocation of new investable cash flows more sensitive to expected yields (and other influences) than the re-allocation of existing holdings.[8] The specification of the two private borrower groups' respective supplies of corporate bonds follows from an analogous treatment of the optimal choice of liabilities to finance a given cumulated external deficit. For consistency with the 1978 version of the MPS model, all equations are estimated using quarterly data to 1976.

With the addition of the structural models of the corporate and government bond markets, the altered MPS model includes an explicit representation, as in (1)–(3) above, of the markets for six assets: money (or reserves, depending on the representation of monetary policy), four maturity classes of government securities,

[6] See Roley (1980, 1982) for details of the estimation. The model takes as exogenous the net bond purchases of all investors other than those noted above. During the sample period used to estimate the model, the explicitly modelled investors accounted for about 55% of all holdings of US Treasury securities; the Federal Reserve System (24%) and foreign investors (18%) accounted for most of the remainder.

[7] The particular yields corresponding to the four maturity ranges are the Treasury yields on 3-mo. bills and bonds in the 3–5 yr, 6–8 yr, and 10-yr-and-over groups.

[8] See Friedman (1977a) for the development of the optimal marginal adjustment model used in all of the corporate bond demand and supply equations and most of the government security demand equations, and Roley (1980) for the development of a more general alternative used in some of the government security demands (especially those for commercial banks).

128   *Benjamin M. Friedman*

and corporate bonds. This system of six asset market equilibrium
conditions is sufficient to determine the yields on five assets,
given that of the sixth. With the yield on money (or reserves)
fixed at zero, therefore, the model determines the yields on the
remaining five assets that are interest-bearing in the conventional
sense.[9] The yields on other assets that appear in the MPS model
follow in the usual way from the original model's term-structure-
like equations, linking each to one of the five structurally determined yields. The Treasury bill rate, for example, directly
determines the commercial paper rate, while the corporate bond
rate directly determines the mortgage rate and the dividend price
yield.[10]

In all other respects, the model underlying the simulations
reported in Chapter 13 is identical to the familiar MPS model.
Because of the richer treatment of the determination of relative
interest rates, however, the altered model (unlike the original)
admits analysis of the effects of debt management policy.

[9] If the estimation of the six sets of asset demand (and supply) equations imposed the full set of balance sheet constraints for all investors (and private-sector
borrowers), solving the model would involve simply deleting the equilibrium
condition for any one market chosen arbitrarily. Imposing these constraints,
however, would have required also re-estimating the MPS model's aggregate
money demand equation. In fact, the constraints are not fully imposed, and hence
the model is overdetermined. In the simulations reported in Ch. 13 below, the
composition of the Federal Reserve's portfolio is adjusted in each period so as to
render consistent the Treasury bill rate proximately determined in the money
market and the Treasury bill rate proximately determined in the shortest maturity
sub-market of the government securities market model (see the discussion in
Ch. 13). Analogous simulations, based on an alternative solution procedure in
which the Treasury bill rate is proximately determined in the government securities market model and the MPS model's money demand equation is deleted,
differ in some specifics but yield the same overall results.

[10] At least in principle, a fully comprehensive demand–supply model of all
asset markets would be preferable. For efforts along these lines see Bosworth
and Duesenberry (1973), Hendershott (1977), and Backus *et al.* (1980). None of
these models, however, distinguished between government and corporate bonds
and among maturity classes of government securities as in the model used here.

13

Empirical Assessment of the Effects of Debt Management Policies

Table 13.1 summarizes the results of simulations of the combined MPS–corporate–government–bond–market model for two different debt management actions: first, a sustained shift in Treasury financing to emphasize new issues of short- instead of long-term securities and, second, a one-year programme to shorten the maturity structure of the outstanding Treasury debt by issuing short- and repurchasing long-term securities.[1] (The simulated effects of debt management actions in the model are sufficiently symmetrical that there is no need to show results of analogous actions to lengthen the debt.) The simulation period in both cases is the ten-quarter interval spanning 1974(4)–1977(1).[2]

In a partial equilibrium analysis of the asset markets like that in Chapter 11, specifying changes in policy-determined supplies of outside assets ($dS = -dL$, for example) is straightforward. By contrast, in a general equilibrium context both fiscal and monetary policy have direct implications for the supply of government securities. For example, if a change in relative asset returns arising from a debt management action stimulates overall economic activity, it will also raise tax revenues and reduce transfers, and in the absence of offsetting increases in government purchases will reduce the government deficit (or increase

---

[1] Because the US Treasury does not ordinarily repurchase its outstanding long-term bonds, it is perhaps easiest to think of such a one-year programme as carried out by Federal Reserve open market operations.

[2] Carrying out the investigation under conditions of underutilized resources in the economy is probably best because of familiar concerns about the underlying MPS model's representation of economic behaviour near full employment. The simulation period chosen here, which falls just at the end of the sample period used to estimate the MPS model, was also interesting in terms of the historical debt management policy. Continuing either simulation beyond ten quarters or so would not be instructive because thereafter the MPS model's well-known long-run oscillatory behaviour begins to dominate; the resulting reversal of effects is already somewhat apparent in Fig. 13.1.

TABLE 13.1 *Simulated effects of two debt management actions*

Variable	Historical mean (1)	$250 m. shift from long to short all 10 quarters		$1 b. shift from long to short first 4 quarters only	
		Simulated mean (2)	Difference from historical (3)	Simulated mean (4)	Difference from historical (5)
r_{TB}	5.50	5.68	0.18	6.17	0.67
r_{35}	7.22	7.29	0.07	7.60	0.38
r_{68}	7.59	7.38	−0.21	7.05	−0.55
r_{TL}	6.90	6.66	−0.25	6.35	−0.55
r_{CB}	8.61	8.49	−0.12	8.41	−0.20
r_{DP}	4.18	4.13	−0.05	4.06	−0.11
X	1241.8	1246.3	4.6	1255.7	13.9
IP	37.0	37.4	0.3	38.1	1.0
IE	79.6	80.2	0.6	81.6	2.0
IH	43.8	45.0	1.1	46.9	3.1
C	798.9	801.1	2.2	805.5	6.6

CUR	75.2	75.3	0.1	0.2
RNB	34.7	34.5	−0.2	−0.7
DF	57.5	55.4	−2.1	−6.5
S	954.8	973.6	18.8	43.1
PRO	111.2	113.2	1.9	5.3

Variable symbols:

r_{TB} = 3-mo. Treasury bill yield (%)
r_{35} = 3–5-yr Treasury security yield (%)
r_{68} = 6–8-yr Treasury security yield (%)
r_{TL} = 10-yr-and-over Treasury security yield (%)
r_{CB} = corporate bond yield (%)
r_{DP} = dividend price yield (%)
X = real GNP (1972 $b)
IP = real investment in plant (1972 $b)
IE = real investment in equipment (1972 $b)
IH = real residential investment (1972 $b)
C = real consumer expenditures (1972 $b)
CUR = currency outside banks ($b)
RNB = non-borrowed reserves ($b)
DF = federal government deficit ($b)
S = market value of common stock ($b)
PRO = corporate profits ($b)

132   *Benjamin M. Friedman*

the surplus). As time passes, therefore, the total amount of government securities outstanding will be less than it would have been otherwise, and it becomes necessary to make some ahistorical assumption about the composition of Treasury financing. The simulations reported in Table 13.1 are based on the assumption that real government purchases are fixed, and that in this situation the induced reduction of debt in each maturity class is proportional to the share of that class in the total Treasury debt outstanding; that is, after the deliberate debt management policy action, the Treasury finances changes from the historical outstanding debt so as not to alter the maturity structure still further.[3]

Similarly, if monetary policy fixes the growth rate of either bank reserves or any given monetary aggregate, the total amount of government securities held by all private investors will decline over time as the central bank conducts the open market operations needed to accommodate the public's greater demand for currency associated with a greater level of economic activity. If monetary policy fixes the growth rate of a monetary aggregate, then still further induced changes in the central bank's holdings will follow as it accommodates the banking system's changing demand for non-borrowed reserves arising from the public's shifting preferences for different kinds of deposits bearing different reserve requirements, as well as from any changes in banks' aggregate net free reserve position. Hence some ahistorical assumption about the composition of the central bank's portfolio is also necessary. The simulations reported in Table 13.1 are based on the assumption that the Federal Reserve System fixes the growth rate of the M1 money stock; that it buys or sells the amount of Treasury bills required to render consistent the values of the Treasury bill rate determined in the money market and in the shortest maturity sub-market of the government securities market model; and that, for the incremental induced changes in the size of its portfolio, it buys or sells the other three maturity

---

[3] With government purchases fixed in real terms, the rising price level offsets part of the increased tax revenues. This effect is small in a 10-quarter simulation, however, so that most of the rise in revenues simply reduces the deficit. It is perhaps useful to note explicitly that this way of treating the financing of the induced reduction in government debt outstanding assumes that the Treasury is not pursuing a policy of minimizing interest costs (at least, not at this particular margin).

Debt Management Policy and Interest Rates   133

classes of government securities together in proportion to their respective total amounts outstanding (so that in this respect it acts analogously to the Treasury's financing of incremental induced deficits or surpluses).[4]

Apart from the assumed changes in debt management policy that are the primary focus of attention, augmented by these additional technical assumptions about the securities transactions endogenous to fiscal and monetary policies, the simulations reported in the table rely on historical values of all exogenous variables. Moreover, each equation in the model is adjusted by adding back the associated historical single-equation residuals so that, given the historical values for all exogenous variables (including supplies of each maturity class of Treasury securities), the model would exactly reproduce the historical values shown in column (1) of the table. The differences between these historical values and the simulated values shown in the table's remaining columns are therefore attributable entirely to the effect of the specified debt management actions, rather than to any underlying inability of the model to reproduce the observed historical record.

Columns (2) and (3) of the table summarize the results of a simulation of the model in which, in each quarter, the Treasury issues $250 million more short-term and $250 million less long-term securities (before adjustment for induced changes in the federal deficit). The historical amounts of short- and long-term Treasury securities outstanding as of 31 March 1977 were $144.5 billion and $20.1 billion, respectively, so that a change of this magnitude in debt management, even cumulated over ten quarters, is small compared with the former but substantial compared with the latter. Because of the Treasury's policy shift to lengthening the debt after 1975, in conjunction with the larger federal deficit in the wake of the 1973–5 recession, the historical amount of long-term Treasury securities outstanding increased by $7.3 billion during these ten quarters. In the simulation the increase is only $4.5 billion.[5] Column (2) shows ten-quarter

---

[4] See again fn. 9 of Ch. 12. Here, too, the assumption is that the Federal Reserve does not act to maximize interest earnings on its portfolio.

[5] The outstanding supply of long-term Treasury securities at the end of the ten quarters in the simulation is lower than the historical by $2.88b instead of $2.50b (10 quarters times $250m per quarter) because of the smaller total volume of

134   *Benjamin M. Friedman*

simulated means for selected financial and non-financial variables, and column (3) shows the respective differences between these simulated means and the historical means shown in column (1).

The simulated effect of this change in outside asset supplies on the Treasury yield curve, shown in the first four rows of the table, corresponds to standard presumptions based on the theory of portfolio behaviour and interest rate determination outlined in Chapter 11. The Treasury bill rate rises and the long-term bond rate falls, compared with the corresponding historical values, and the relative movements in the rates on the two intermediate maturity classes lie between the two extremes.[6] Similarly, the effects on private asset yields, shown in the next two rows, correspond to familiar notions of relative asset substitutabilities. The average relative yield decline for corporate bonds is only half that for long-term government bonds, while the relative decline for the equity yield is smaller still.

The next five rows of the table indicate the effects of this debt management action on both the level and the composition of real economic activity. Real output is greater than the historical by nearly 0.5 per cent on average. Moreover, because of the sensitivity of investment to cost-of-capital factors, fixed capital formation accounts for a disproportionate amount of the increase. The three categories of fixed investment, which together comprised only one-eighth of total spending, account for nearly one-half of the simulated increase over the historical. Hence the results support Tobin's (1963) conclusion that shortening the maturity structure (in contrast to the US Treasury's recent debt management policy) would enhance the economy's rate of capital formation.

The bottom five rows of the table present values for additional financial variables that are useful for understanding the structure of the simulation. The induced effects on the Federal Reserve's portfolio are small, given the assumption that it fixes the M1 money stock. Small open market purchases are necessary to accommodate the public's increased demand for currency, but

---

Treasury financing arising from the induced rise in tax revenues and fall in transfers.

[6] An interesting result is the apparent strong substitutability between securities in the third and fourth maturity classes. This result appears even more strongly in the second simulation, reported in columns (4) and (5).

Debt Management Policy and Interest Rates   135

slightly larger open market sales are necessary to accommodate banks' reduced demand for non-borrowed reserves arising from smaller holdings of demand deposits and larger discount window borrowings.[7] By contrast, the induced effects on the total amount of Treasury securities outstanding are more substantial. A large increase in tax revenues and a small decline in transfer payments reduce the federal deficit and hence the required volume of Treasury financing.[8] Finally, an apparent puzzle in the simulation is that the increase in consumer spending is surprisingly large compared with the small decline in the dividend price yield, given the underlying MPS model's reliance on the life-cycle model of consumption. The explanation, as shown in the table, is that equity prices do rise substantially, in large part because of an increase in corporate profits (which in turn raises dividends).

Figure 13.1 provides further information about these results by plotting the historical (solid-line) and simulated (broken-line) paths, quarter by quarter for all ten quarters, for six key variables. The increase in the Treasury bill rate, compared with the historical, takes place gradually. By contrast, the relative decline in the long-term Treasury bond rate occurs within two quarters, while the relative decline in the corporate bond rate also occurs within two quarters but then almost disappears after another six. The relative increase in equity prices reaches its peak after six quarters. The stimulative effect of these financial developments on real output nears its peak by the sixth quarter, but in fact continues to build through the ninth quarter and then declines only negligibly in the tenth. The stimulative effect on real capital formation follows a roughly similar path, with a peak in the eighth quarter and a negligible decline thereafter.

Columns (4) and (5) in Table 13.1 summarize the results of a simulation of the model in which, in each of the first four quarters, the Treasury issues $1 billion more short-term securities and repurchases enough long-term securities to reduce the net

---

[7] The public's shift from demand deposits to currency, within a fixed M1 total, is the conventional result associated with greater real income and higher short-term interest rates. The small increase in borrowed reserves occurs mostly because the simulation holds the discount rate fixed despite the rise in short-term market rates.

[8] The reduction of the deficit is surprisingly large for the associated increase in income, even after allowance for the difference between nominal and real magnitudes.

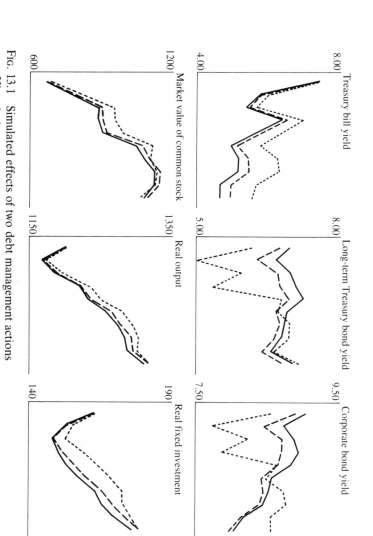

FIG. 13.1 Simulated effects of two debt management actions
— Historical
--- Simulated action: $250m shift from long to short, all 10 quarters
······ Simulated action: $1b shift from long to short, first 4 quarters only

Debt Management Policy and Interest Rates 137

flow supply of the latter by $1 billion relative to the historical (before adjustment for induced changes in the federal deficit). For the remaining six quarters of the simulation, debt issues follow the historical proportions. Such a one-year bill–bond swap programme would have represented a major debt management operation, especially in the context of the limited size of the long-term Treasury bond market at the time. From $12.7 billion as of 30 September 1974, the outstanding amount of long-term Treasury securities in the simulation falls to $10.3 billion a year later, in contrast to the historical rise to $14.5 billion.[9] Column (4) shows the simulated ten-quarter means, and column (5) shows the respective differences between these simulated means and the historical means in column (1).

The average effects of this one-year swap on the structure of asset yields are analogous to, but for each yield greater than, the average effects of the sustained change in the maturity of new issues studied in the first simulation. Within the Treasury yield curve, once again the bill rate rises and the long-term rate falls compared with the corresponding historical levels. The rate for the second maturity class again moves in the same direction as that for the first, and again by about half as much, while this time the rate for the third maturity class moves exactly as much as that for the fourth. The relative declines in the corporate bond yield and the dividend price yield are of about the same proportion, when compared to the relative decline in the long-term Treasury yield, as in the first simulation.

The associated effects on real economic activity are also similar, though larger throughout. Real output is greater by about 1 per cent on average, with fixed capital formation again accounting for nearly half of the increase. The Federal Reserve's holdings of Treasury securities decline more substantially than in the first simulation, almost entirely as a result of the reduced demand for non-borrowed reserves associated with greater discount window borrowings;[10] and the decline in the total amount of Treasury securities outstanding is also more substantial as a result of

---

[9] The difference is $4.22b instead of $4.0b (4 quarters times $1.0b per quarter) because of the smaller total volume of Treasury financing arising from the induced rise in tax revenues and fall in transfers. By the end of the simulation period, the outstanding amount is $5.19b less than the historical.

[10] See again fn. 7.

the stimulation of additional tax revenues. The combination of a steeper relative decline in the dividend price yield and a greater rise in corporate profits (and hence dividends) leads to a much larger rise in equity prices than in the earlier simulation.

Although comparison of the ten-quarter means shown in Table 13.1 suggests that the economic effects of the one-year swap are just an enlarged mirror of the effects of the sustained change in new issue design, the quarter-by-quarter time paths plotted in Fig. 13.1 (the dotted lines) make it clear that this is not so. The most immediate contrast is in the effects on the long-term asset markets—including the long-term Treasury bond yield, the corporate bond yield, and the price of equities. These effects in each case build irregularly during the four quarters in which the debt management action is in progress, but then decline rapidly thereafter and even change sign during the latter part of the simulation.

The increase in real output compared with the historical reaches a peak, equal to nearly 2 per cent of the corresponding historical output, in the sixth quarter. Thereafter it too erodes rapidly, so that by the tenth quarter the respective output paths for the two simulations converge. The peak effect on fixed investment also occurs in the sixth quarter, after which the increase erodes especially rapidly because of the effects of high short-term interest rates on residential construction (which falls below the historical by the end of the simulation period).

The Treasury bill rate is the one variable for which the time path resulting from the one-year swap most nearly resembles that from the sustained debt management change. The increase in the bill rate compared with the historical continues to build almost until the end of the simulation period, by which time the effect on real income has largely eroded.

The contrasts between these two simulations indicate that, in addition to the magnitude, the timing of a debt management action affects its impact on the economy. This result is not surprising in light of the basic model of portfolio adjustment underlying both the corporate and the government bond market models. Because the optimal marginal adjustment model distinguishes between investors' allocation of new cash flows and reallocation of existing holdings (and draws an analogous distinction between private borrowers' financing of new external

Debt Management Policy and Interest Rates

deficits and refinancing of existing liabilities), in the short run financial flow variables matter in addition to stock variables. Hence the size of a debt management action in relation to other flows in the financial system is a key determinant of its effects.

More importantly, however, despite their contrasts, the two sets of results both indicate that debt management actions have effects—on interest rates as well as on non-financial economic activity—that not only are in accordance with familiar theory but also are of a size deserving attention. The finding that these simulated effects are of substantial magnitude conforms to the implications of at least some of the literature of partial equilibrium analysis of asset substitutabilities based on the observed properties of asset returns.

14

Summary and Conclusions

The maturity structure of the US government's outstanding debt has undergone large changes over time, at least in part because of shifts in the Treasury's debt management policy. During much of the first three decades following the Second World War, an emphasis on short-term issues rapidly reduced the debt's average maturity. For a while during the early 1960s and again since 1975, however, the opposite policy has just as rapidly lengthened the average maturity.

Debt management actions do not leave other aspects of economic activity unaffected. Within the financial markets, changes in the relative supplies of outside assets in general alter the structure of expected yields on all assets whether outside or inside, raising the relative yield on the asset with supply increased and on assets closely substitutable for it, and lowering the relative yield on the asset with supply decreased and on its close substitutes. In a general equilibrium context the resulting realignment of asset yields and prices also affects non-financial economic activity.

Simulation experiments indicate that debt management actions of a magnitude comparable to observed changes in US debt management policy have sizeable effects both in the financial markets and more broadly. In particular, a shift from long- to short-term government debt lowers yields on long-term assets (and raises their prices), raises yields on short-term assets, and in the short run stimulates output and spending. Moreover, the stimulus to spending is disproportionately concentrated on fixed investment, so that debt management actions shortening the maturity of the government debt not only increase the economy's output, but also shift the composition of output towards increased capital formation.

15

Comment

STAFFAN VIOTTI

As stated, the objective of Professor Friedman's paper is 'to provide a quantitative assessment of the economic effects of debt management policy'. For this purpose he uses the MIT–Penn–SSRC (MPS) econometric model of the USA with one important modification. The term-structure equation of the MPS model is replaced by a structural model 'representing the determination of interest rates in four separate sub-markets of the US government securities market'. In Chapter 11 this structural model is presented in some detail, followed in Chapter 12 by a presentation of the results of two debt management policy simulations. The first policy experiment is a sustained shift towards short-term instead of long-term financing of a given total government debt; the second experiment performs a similar shortening of the maturity structure but within one year. Chapter 13, finally, contains a detailed presentation of the portfolio model that is claimed to provide structure for the financial sub-model added to the MPS model.

It is difficult to express definite views on whether the policy experiments give a reasonable picture of the possible effects (qualitative or quantitative) of debt management policy. Friedman's presentation of the model is too brief and my own familiarity with the details of the MPS model is too superficial. My suspicion, though, is that, as with most large-scale macro-econometric models, in spite of overwhelming richness in details, what is really driving the model is some relatively simple textbook macro framework of Keynesian flavour. Thus, the experiments might be regarded as an empirical check on the propositions put forward in Friedman's much cited paper on crowding-in and crowding-out (Friedman 1978).

In that paper, some basic results concerning the effects of various government debt policies are derived by extending the

142   *Staffan Viotti*

financial sector of a conventional Keynesian macro model along the lines originally introduced by Tobin (1969). In the Tobin approach, although reference is made to portfolio analysis, no results are explicitly derived from portfolio optimization. As in most (pragmatic) macro modelling among a large number of financial assets, a few aggregate or representative assets are picked out. For each of these, ordinary demand functions are assumed. Some restrictions well known from demand analysis such as adding up restrictions are applied. Others, like gross substitutability and own effects being larger (in absolute magnitude) than cross-effects, are also imposed in qualitative analyses, mainly with reference to the intuitive plausibility of the assumptions.

To derive quantitative results, ordinary econometric methods are used to estimate the parameters of the demand functions. Most large-scale econometric models have some (extended) LM sectors obtained in this fashion. Obviously, this approach does not make use of any results from portfolio analysis and capital asset pricing. Although one (maybe the main) criterion for distinguishing between assets is their degree of risk, this is not explicitly modelled in the Tobin approach. This means that there is a lot of information, for instance the matrix of co-variances between the assets that could be estimated from the time series of realized returns, which is not fully utilized on the estimation process.

In the above paper Friedman tries to put more structure into the financial sector by applying portfolio analysis and asset pricing results. The one-period Markowitz–Sharpe–Lintner framework is used to derive demand relations for the asset aggregates. With supplies exogenously given in equilibrium, these demand functions could be inverted to solve for equilibrium returns. The parameters of these CAPM (capital asset pricing model)-type equations are, among other things, explicitly determined by the covariance matrix.

The important question is whether this is a fruitful way to add structure to this financial sub-sector. I doubt that it is, and the reason is basically that the one-period mean–variance model is ill-suited to deal with portfolios containing interest-bearing assets of various maturities. In the classical Sharpe–Lintner framework, assets are valued according to their covariation with the market

Debt Management Policy and Interest Rates 143

portfolio. In other words, the returns on equilibrium assets are determined exclusively by their contributions to diversification. This may be a reasonable approach when dealing with portfolios consisting only of stocks. In Friedman's case the portfolio consists mainly of interest-bearing assets, and here the one-period portfolio approach probably becomes misleading. Intuitively, we feel that interest-bearing assets are included in portfolios not mainly for diversification, but for hedging purposes. The most obvious example of the latter is the use of default-free bonds as the 'safe' asset for an investor having a specific time-point for liquidating his portfolio. As is well known from intertemporal asset pricing models, the one-period Sharpe–Lintner model provides a good approximation of the more general multi-period case only if consumption and investment opportunities either are not changing over time or are changing in a completely predictable (deterministic) or completely unpredictable manner. Since the term structure of interest rates is one of the most important determinants of investment opportunities, this implies that the one-period portfolio model is a good approximation if the term structure is changing over time in a deterministic way or if interest rates move completely randomly. That the first assumption is no good approximation of reality is obvious for anyone who is actually trying to make forecasts of future interest rate movements. The second assumption is even more unrealistic, since it implies that the level of interest rates today has nothing to do with the level of interest rates tomorrow. Interest rates over time are independent drawings from a (possibly changing) probability distribution.

It seems preferable to have less structure than more, if the added structure obviously goes against empirical evidence. My conclusion therefore is that adding structure derived from one-period portfolio optimization to a financial sector consisting mainly of interest-bearing assets is not very meaningful.

What, then, are the alternative ways to model the financial sector? The most ambitious approach would be to draw on modern theoretical work on multi-period portfolio analysis and asset pricing and its application for term structure modelling. The problem is, of course, that empirical work utilizing this highly complex framework has just begun. (For an example of such work, see Brown and Dybvig 1986.)

144   *Staffan Viotti*

While awaiting interesting results from this ongoing work in theoretical and empirical finance, in my view the simple Tobin framework will do. That is, asset demand functions are assumed with no explicit reference to portfolio theory. Note, however, that a lot of structure is implicit in Tobin-type models. First, the choice of aggregate or representative assets has such implications. For instance, modelling bond markets with a single bond or with a long and a short bond seems closely related to the one- or two-parameter representations of the term structure being used in duration analysis. Second, the identification of sector demand functions like in Friedman's paper can be interpreted as an assumption about different 'habitat' preferences among investor groups. The distinction between demands for risky assets in the banking and insurance sectors is probably motivated not by differing diversification needs but by different preferred habitats.

Analyses of government debt policies based on simulation experiments with econometric models including a Tobin-type financial sector are of course very much open to Lucas critique arguments. The estimated parameters should not be given structural interpretations. But cautiously interpreted, I think these experiments can be a useful first check on the possible effects of debt policies, if any. If they repeatedly show similar patterns, we may be provided with some kind of stylized facts to take as a point of departure for a deeper analysis of the mechanisms through which government debt policy might affect the economy. As pointed out by Agell and Persson in their paper in this volume, a natural reference model is given by Modigliani–Miller-type arguments. In a world with perfect and complete capital markets, government debt policy would not matter. If the data indicate that debt policy does matter, which for instance Friedman's experiments seem to do, the next step must be to try to build models that focus on the reasons for these deviations from the reference model. Some such models already exist based on intergeneral arguments *à la* Barro. Personally, I think that this approach is of second-order importance to explain the possible effects of government debt policy. A more careful analysis of the financial intermediation process is probably the clue towards a better understanding of how debt policy works. Empirical analysis in this area has been pioneered by, among others, Benjamin Friedman in his studies of credit flows. The new, more analytical

Debt Management Policy and Interest Rates

approach to the study of financial intermediaries which has been developing on the 1980s should give important inputs into the analysis of government debt policy.

References to Part II

Backus, D., Brainard, W. C., Smith, G. and Tobin, J. (1980), 'A Model of US Financial and Nonfinancial Behavior', *Journal of Money, Credit and Banking*, 12: 259–93.

Blanchard, O.-J. and Plantes, M. K. (1977), 'A Note on Gross Substitutability of Financial Assets', *Econometrica*, 45: 769–71.

Bosworth, B. and Duesenberry, J. S. (1973), 'A Flow of Funds Model and Its Implications', *Issues in Federal Debt Management*. Boston: Federal Reserve Bank of Boston.

Brown, S. and Dybvig, P. (1986), 'The Empirical Implications of the Cox, Ingersoll and Ross Theory of the Term Structure of Interest Rates', *Journal of Finance* 41(3): 617–32.

Frankel, J. A. (1985), 'Portfolio Crowding-Out Empirically Estimated', *Quarterly Journal of Economics*, C (suppl.): 1041–65.

Friedman, B. M. (1977a), 'Financial Flow Variables and the Short-Run Determination of Long-Term Interest Rates', *Journal of Political Economy*, 85: 661–89.

——(1977b), 'The Inefficiency of Short-Run Monetary Targets for Monetary Policy', *Brookings Papers on Economic Activity*, no. 3: 293–335.

——(1978), 'Crowding Out or Crowding In? Economic Consequences of Financing Government Deficits', *Brookings Papers on Economic Activity*, no. 2: 593–641.

——(1979), 'Substitution and Expectation Effects on Long-Term Borrowing Behavior and Long-Term Interest Rates', *Journal of Money, Credit and Banking*, 11: 131–50.

——(1980), 'The Effect of Shifting Wealth Ownership on the Term Structure of Interest Rates: The Case of Pensions', *Quarterly Journal of Economics*, 94: 567–90.

——(1985a), 'The Substitutability of Debt and Equity Securities', in B. M. Friedman (ed.), *Corporate Capital Structures in the United States*. Chicago: University of Chicago Press.

——(1985b), 'Crowding Out or Crowding In: Evidence on Debt–Equity Substitutability'. Mimeo, National Bureau of Economic Research.

—— and Roley, V. V. (1987), 'Aspects of Investors' Behavior Under Risk', in G. R. Feiwel (ed.), *Arrow and the Ascent of Modern Economic Theory*. New York: New York University Press.

Debt Management Policy and Interest Rates 147

Hendershott, P. H. (1977), *Understanding Capital Markets*, i: *A Flow-of-Funds Model*. Lexington, D. C. Heath.

Lintner, J. (1969), 'The Aggregation of Investors' Diverse Judgements and Preferences in Purely Competitive Security Markets', *Journal of Financial and Quantitative Analysis*, 4: 347–400.

Modigliani, F. and Shiller, R. J. (1973), 'Inflation, Rational Expectations, and the Term Structure of Interest Rates', *Econometrica*, 4: 12–43.

—— and Sutch, R. (1966), 'Innovations in Interest Rate Policy', *American Economic Review*, 56: 178–97.

—— and —— (1967), 'Debt Management and the Term Structure of Interest Rates: An Empirical Analysis of Recent Experience', *Journal of Political Economy*, 85: 569–89.

Okun, A. M. (1963), 'Monetary Policy, Debt Management, and Interest Rates: A Quantitative Appraisal', in Commission on Money and Credit, *Stabilization Policies*. Englewood Cliffs, NJ: Prentice-Hall.

Roley, V. V. (1979), 'A Theory of Federal Debt Management', *American Economic Review*, 69: 915–25.

—— (1980), 'A Disaggregated Structural Model of the Treasury Securities, Corporate Bond, and Equity Markets: Estimation and Simulation Results'. Mimeo, National Bureau of Economic Research.

—— (1982), 'The Effect of Federal Debt Management Policy on Corporate Bond and Equity Yields', *Quarterly Journal of Economics*, 97: 645–68.

—— (1983), 'Symmetry Restrictions in a System of Financial Asset Demands: Theoretical and Empirical Results', *Review of Economics and Statistics*, 65: 124–30.

Tobin, J. (1963), 'An Essay on Principles of Debt Management', in Commission on Money and Credit, *Fiscal and Debt Management Policies*. Englewood Cliffs, NJ: Prentice-Hall.

—— (1969), 'A General Equilibrium Approach To Monetary Theory', *Journal of Money, Credit and Banking*, 1: 15–29.

Index

Agell, J. 1, 2, 3, 72 n., 80 n., 81, 82, 84, 87, 88, 89, 91, 92, 93, 94, 95, 96, 97, 98, 99, 100, 101, 102, 144
agents 18, 19, 25, 31, 45–52, 54
aggregation 2, 10, 39, 52–8, 79, 95–6
Ahn, C. M. 101 n.
Akgiray, V. 101 n.
amortization 14, 23
ARCH method 86
Argentina 30 n.
ARMA process 43
assets:
  corporate 86
  equilibrium 143
  foreign 27, 28, 100
  holding period 54
  inside 119, 120
  interest-bearing 76, 128, 143
  non-marketable 80
  numeraire 71
  outside 119
  outstanding 76
  pricing 8, 12, 143
  private 21
  real 25, 41, 42
  riskless 63
  risky 72 n., 94, 115 n., 116 n.;
    assessing 11; covariance between 36; demand for 34 n., 144
  safe 47, 143
  underlying 60, 61
Auerbach, A. J. 30
Australia 7 n.

Backus, D. 82 n., 128 n.
Ball, C. A. 101 n.
'bangs' 102
bank loans 120
Barro, R. J. 19, 144
Bergstrom, A. R. 54 n., 81 n.
Bergström, V. 3
Berndt, E. R. 54 n.
Bernheim, B. D. 25 n., 99 n.
Black, F. 60, 61
Blanchard, O. 72 n., 118 n.

Blume, M. E. 33 n., 37 n.
Bodie, Z. 41, 84 n.
Bollerslev, T. 52, 59, 86
bonds 16–17, 18, 55, 69, 111, 120
  corporate 125, 126, 127, 128, 134, 135
  default-free 143
  demand for 13
  discount 14, 20, 21, 67
  indexed 19, 25, 30 n., 31
  local 76
  newly issued, public forced to buy 30
  private 21, 22, 80
  Treasury 61
  utility 126 n.
  see also long-term bonds; short-term bonds; yields
Booth, G. G. 101 n.
borrowing 8, 18, 22, 26, 29–30, 121
  discount window 135, 137
  domestic and foreign 17
  government spending financed by 13, 14, 19, 20
  individuals and firms 119
  minimizing costs 9
  shorter-term 111
  strategies 15
Bosworth, B. 128 n.
Box, G. E. P. 43
Brainard, W. 32, 82 n.
Brazil 30 n.
Brown, S. 143
Brownlee, O. H. 9 n., 32
budget constraints 14, 19, 20, 22, 26
  household 24
  individual 34
business cycle 42

Canada 95 n., 100
capital 28, 81, 82, 118–19, 120–1
  allocation 29
  asset pricing 142
  cost 47, 99, 125, 134
  fixed 134, 137

capital (*cont.*)
  formation 134, 135, 140
  gains 39, 44, 91
  loss 30, 39
  mobility 89, 100
CAPM-type equations 142
CASE method 85, 86
cash flows 14, 15, 138
central banks 9, 12, 16, 132
Chamley, C. 25
Chan, L. K. C. 20 n., 22 n., 24
Chicago 61
Chile 30 n.
Chouraqui, J.-C. 7 n., 8
Clark, P. K. 101 n.
coefficients 43, 85, 87
  correlation 40, 41, 47, 55, 57, 63
  risk aversion 83, 86, 87, 91, 115
  substitutability 82
commercial paper rate 128
consumer durables 40, 80, 89, 100, 125
consumer price index 40, 42
consumption 14, 21, 22, 23, 24–5, 143
  intertemporal 19, 24
  life-cycle function 99, 135
  lifetime, expected utility 20
  non-durable 125
corporate equity 36, 51, 62, 70, 74, 76
  yields 50, 78
corporate stock 42, 45
  covariance matrix 39, 40, 44
  yields 41, 46, 47, 56, 57
coupons 67, 75, 111
covariance 2, 54, 72, 74, 76
  conditional 46, 49, 50, 56
  unconditional 44, 49
covariance matrices 10, 34–45, 47, 51,
  52, 66, 68–70, 79, 118 n.
  alternative approach to 59–65
  CAPM-type equations determined
    by 142
covered interest differential 89, 90
Cox, J. C. 16 n., 80, 101 n.
credit 29, 120, 126
currency 131, 132, 135 n.

deficits 1, 111, 131, 135, 139
  bond-financed 17
  fiscal 86
  induced changes 133, 137
  persistent 7
  reducing 129, 135

De Long, J. B. 95 n.
demand 33–4, 83, 113, 124, 142
  aggregate 24, 27, 50
  behaviour 92, 93
  bonds 13
  deposits 135
  equity 17
  expected returns 60
  functions 70 n., 144
  household 24
  industry investment 68 n.
  inferred 98
  money 116 n., 121, 123
demand system 37, 99, 115 n., 117
deposits 125 n., 132
depreciation method 51
derivatives 58, 77, 99–100, 118
  comparative-static 69
  general equilibrium partial 73
  policy 47, 57, 63–4, 65, 75, 76
  unconditional 49
distribution:
  joint asset return 115
  normal 102
  Paretian 101
  probability 143
  stochastic 101
dividends 39, 75, 123, 135, 138
Duesenberry, J. S. 111 n., 128 n.
Dybvig, P. 143

Engel, C. 84, 86
Engle, R. F. 52, 59
equities 31, 79, 121, 123, 125 n.
  excess returns on 96, 98–9
  see also corporate equity
estimation 82, 100, 126–7 nn., 128 n.,
  142
  constrained asset share 84–91
  GARCH 43 n., 51–2, 59
Estrella, A. 111 n.
exchange rate 28
expectations 60, 67, 82, 85, 86, 97
expected returns 34 n., 36, 60, 66
  absolute 71
  changes in supplies 89
  determination 71, 116
  real 33
  regression 82–3
  short-term debt 50
expenditure 15, 17, 78, 121
  consumer 125, 131

Index

decisions of firms 32, 76
ways of financing 13–14, 19, 20, 27

face values 69
Fair, R. 82 n.
Fama, E. F. 101
'fat tails' problem 96, 101
Federal Reserve 45 n., 123, 127 n.,
   128 n., 129 n., 132, 133 n., 134, 137
Feinstone, L. J. 101 n.
fiscal policy 10, 11, 14, 92, 133
   debt management and 26; mixed
   18; separate from 13, 15, 17
Fischer, S. 31 n., 55 n., 96 n.
forecasting errors 82, 85
forecasts 43
France 89
Frankel, J. 2, 7 n., 18, 38 n., 41, 42,
   49, 50, 52 n., 60, 84, 86, 87, 90,
   99 n., 100, 124 n.
Friedman, B. M. 1–2, 3, 7 n., 10 n.,
   17, 18, 19, 32, 33, 38, 45, 50, 51,
   52, 60, 66, 76 n., 82 n., 83 n., 84 n.,
   86 n., 87, 88, 94 n., 95, 96, 97, 100,
   101, 113, 115 n., 116 n., 117 n.,
   118 n., 123 n., 124, 126 n., 127 n.,
   141, 142, 143, 144
Friend, I. 33 n., 37 n.
Froot, K. 86 n.
futures 61

GARCH approach 43 n., 51–2, 59
Germany 9, 89, 100
GNP 1, 7, 8, 131
Grabbe, O. 111 n.
Grossman, S. 37 n., 54 n.

hedging 143
Hendershott, P. H. 128 n.
high-inflation countries 30 n.
holding period 54, 55
human capital 21, 80

IMF (International Monetary
   Fund) 39, 40, 44
income 116 n., 121, 135 n.
   disposable 21
   real 138
   state-dependent capital 23
inflation 63 n., 86 n., 94, 127
   future 44
   risk 33, 47

uncertainty 70, 94
interest 14, 27, 133
interest rates 29, 30, 55 n., 85, 115–22,
   143
   active debt management, effect on
   3, 139
   determination 113, 134, 141
   long-term 45, 86 n.
   model of, and economic
   activity 123–8
   risk-free 60, 61
   short-term, higher 135 n.
investment 99, 120, 131
   corporate 31, 50
   fixed 134
   industrial 49
   opportunities 33, 143
   private 17, 36
Israel 30 n.
Italy 89

Jacobian 117
Japan 1, 7, 9, 95 n., 100
Jarrow, R. A. 101 n.
Jenkins, G. M. 43
Johnson, D. 111 n.
Jorgenson, D. W. 27, 99

Keynesian models 16, 141, 142
King, M. A. 27, 30
kurtosis 96
Kuttner, K. N. 5, 52, 66, 84 n., 95, 97,
   98

Laibson, D. I. 96, 101
Leape, J. 27
leptokurtosis 96, 101
liabilities 12, 50, 92, 139
   interest-bearing 92
life insurance companies 30 n.
linear stochastic production
   theory 16 n.
Lintner, J. 111 n., 116 n., 142, 143
LM sectors 142
long-term bonds 12, 32, 36, 61, 75, 76
   aggregate household demand 24
   amortization 23
   covariance matrix of real capital
   gain 39, 40, 41, 42, 44
   fall in rate 134
   and short-term bonds 22, 79, 117,
   124; substitution 71, 78, 118

152   *Index*

long-term bonds (*cont.*)
  variance 45, 62, 87
  yields 48; conditional correlation
    coefficients between 57;
    covariances between 46, 47, 56;
    expected 49–50
Lucas, R. E. 16 n., 26, 98, 99

M1 money stocks 16, 17, 132, 134,
  135 n.
McDonald, R. L. 30
Malkiel, B. 82 n.
market-clearing 116, 119, 121
  equilibrium condition 127
  structure 100, 113, 115, 124
markets 26, 51, 78, 120, 128
  bond 113
  capital 25, 89
  failures 31
  international financial 89
  non-financial 122
  rumours 65
  second-hand 8
Markowitz, H. M. 32, 142
Masson, P. 83 n.
matrices:
  coefficient 85
  substitutability 81, 82
  variance-covariance 83, 85, 86, 115
  see also covariance matrices
Mattione, R. 111 n.
maturity composition 23, 31, 50, 78
  changes in 22, 25, 32, 81
Mayers, D. 80
mean-variance 32, 38, 88
medium-term note issues 111
Mehra, R. 88 n.
Merton, R. C. 33, 101 n.
Miller, M. H. 18 n., 19, 25, 144
Modigliani, F. 18 n., 19, 25, 82 n.,
  111 n., 123, 144
monetary policy 11, 14, 59, 125, 132,
  133
  debt management and 26, 84;
    distinction between 10, 13, 15
  demand for money and 123
money 12, 13, 19, 25, 118, 127
  creation 13, 14, 20 n.
  inside 120
  stocks 16, 17, 132, 134, 135 n.
Moody's Investor Service 126 n.
mortgages 120, 125

MPS econometric model 3, 98, 113–
  14, 126 n., 128, 129
  altered 125, 127, 141
Musgrave, R. A. 9 n., 30, 32

neutrality 20, 24
neutrality theorems 10
non-financial economic activity 98–
  101, 113, 114, 125, 126, 139
non-stationarity 51
Norway 7 n.

OECD countries 7, 8, 9
Okun, A. 9 n., 32, 123 n.
Oldfield, G. S. 101 n.
'Operation Twist' 123
optimal allocation 115–16
optimization 14, 87, 91
options 61, 62, 63, 65
  pricing 11, 60, 61
output 137, 138, 140

partial equilibrium analysis 124–5,
  129, 139
pay-off structure 21, 22
Peled, D. 25
Pennacci, G. 96 n.
Persson, M. 1, 2, 16 n., 72 n., 80 n.,
  81, 82, 84, 87, 88, 89, 91, 92, 93,
  94, 95, 96, 97, 98, 99, 100, 101,
  102, 144
Pesando, J. 111 n.
Phelps, E. S. 14
Pindyck, R. 37 n.
Plantes, M. 72 n., 118 n.
Poisson process 101, 102
Polemarchakis, H. 25
Pond, L. 101 n.
portfolios 43, 76, 80, 88, 138, 141
  allocations 115, 127
  assets as close substitutes 19, 38
  balance: approach 32–7;
    equation 83; model 2, 66
  behaviour 30, 124, 134
  choice 7, 26, 45, 78
  considerations 100
  crowding-in 17, 72, 74, 141; and
    crowding-out 18, 36, 41, 50, 57,
    81
  demands 55
  diversification 143, 144;
    optimal 83–4, 85, 86, 87, 91

Index

Federal Reserve 128 n., 132, 133
holdings 74, 75
household 121
minimum-variance 116 n.
optimization 142, 143
private-sector 113
readjustments 17
reallocations 54
rest of the world 89
strategy 23, 24, 25
Poterba, J. M. 95 n.
predictable cycle 42
preferences 116, 119, 144
preferred habitat model 123
Prescott, E. C. 88 n.
price-earnings ratio 99
prices 80, 88, 98, 114–22
current 11, 21, 79, 89, 91
endogenous 2, 66–77, 79
equilibrium 23, 24
equity 135, 138
future 44 n., 91
market-clearing 22
options 60, 63, 86
privatization 13
profits 30, 99, 131, 135, 138
Purvis, D. 82 n.

random disturbance 42, 43
rates of return 71, 81, 82
expected 84, 86 n., 91
real estate 80, 86, 89
refinancing 111
regressions 82–3, 84
replacement cost 120
reserves 128, 131, 132, 135
resource allocation 22, 27
returns 37, 75, 83, 101–2, 142
equilibrium 11, 93
equity 94, 98
excess 94, 95, 96, 98, 100
observed 95, 98, 139
stochastic structure/environment 93, 97
uncertainty of 55 n.
see also expected returns; rates of return
revaluation 76
revenue 14, 20, 22
Ricardian equivalence theorems 15, 18, 19, 79, 98, 99
insights 25

irrelevance literature 26
Riksgëldskontoret 1
risk 10, 39, 78, 85, 88 n., 93
and return 22, 100
risk aversion 33, 35, 41, 70 n., 87, 91
relative 37, 73, 74, 83, 86, 87, 88;
constant 115
Rodriguez, A. 86
Roley, V. V. 32, 33, 41, 50, 52 n., 67, 72 n., 83 n., 84 n., 111 n., 113, 115 n., 117 n., 124, 127 n.
Rolph, E. R. 9 n., 32
Rosenfeld, E. R. 101 n.
Ross, S. A. 101 n.

savings 21, 24, 80, 122
Scholes, M. 60
Scott, I. O. 9 n., 32
securities 100, 113, 128 n., 133, 137
callable 112
non-marketable 12
private 20, 21
short-term 3, 129
see also bonds; equities; Treasury bills
serial correlation 93–5, 102
shares 44, 55, 76, 80
Shiller, R. 37 n., 123
short-term bonds 32–3, 40, 42, 44, 47
aggregate household demand 24
real yields: conditional coefficients of correlation 57; covariance 46, 56; variance 39
see also long-term bonds
short-term financing 49
skewness 96
Smith, G. 82 n.
social costs 29
solvency constraint 15
spending, see expenditure
stabilization policies 9, 27, 51, 68 n.
Standard & Poor's 500 Index 39, 61
stationary autoregressive system 51
Stiglitz, J. E. 22 n., 24
stochastic processes 19, 43 n., 51, 52, 53
stock market crash (1987) 61, 63, 102
Stockholm 81 n.
Stokey, N. L. 16 n.
structural models 26, 113
substitutabilities 81–2, 98, 117, 124, 142

Index

substitutabilities (*cont.*)
  different 38
  high 91
  imperfect 113
  key 10
  pairwise 116
  partial equilibrium analysis 139
  perfect 83
  relative 118, 134
substitutes 2, 31, 38, 72 n., 74, 118
  perfect 21, 22
substitution 78, 116
Summers, L. H. 95 n.
supply 36, 68, 72, 98, 116, 123
  aggregate 50
  bonds 48, 71 n.
  effects 70
  equilibrium 142
  exogenous 35
  outside 134
surpluses 132, 133
Sutch, R. 82 n., 123
Sweden 1, 9, 30, 89, 100, 101
Switzerland 89
symmetric Jacobian 117 n.

tax 13, 19, 22, 27, 32, 78
  borrowing exempted from 29–30
  collections 20
  future 15, 18 n., 25
  increases 18
  inflation 13
  rates 37
  revenues 129, 132 n., 134 n., 135,
    137 n., 138; aggregate 23
  rules 55
  share 24, 25
Taylor expansion 73
Thompson, H. E. 101 n.
time series 43, 54, 55 n., 95, 96, 142
Tobin, J. 9, 17, 19, 26, 27, 29 n., 31,
  68 n., 111 n., 120, 134, 142, 144
Torous, W. N. 101 n.
transaction costs 54, 55, 96, 115, 127
transfers 129, 134 n., 135, 137 n.
Treasury bills 12, 39, 50, 61, 63 n., 132
  rate 128, 134, 135, 138
Tucker, A. L. 101 n.

uncertainty 21, 43, 44, 63, 67, 83
  asset returns 55 n.
  capital gain 39

future purchasing power of debt
  holdings 31
United Kingdom 7, 9, 30 n., 89, 95 n.,
  100
United States 39–41, 100–1, 102
  bond market models 124
  budget deficit 7
  excess returns 95, 96
  fiscal deficits 86
  household wealth 76
  maturity structure of Treasury
    debt 9, 112, 129, 134, 140;
    mean 111, 113
  Reagan tax policy 1
  risky assets 94
  securities market 3; maturity
    classes 126, 127, 128 n., 132,
    134 n., 137
  tax-exempt debt 30
  see also Federal Reserve; MPS
    econometric model
univariate autoregressive process 84 n.
utility 20, 24, 33
  expected intertemporal 88
utility function 88

valuation changes 66, 71, 77
value terms 68, 69, 70, 79
VAR (vector autoregression) 10–11,
  38, 42–4, 50 n., 55, 79, 84, 97
  efficient in explaining inflation 47
  moving-sample 75–6, 78
  reliable 45
variance 2, 43, 55 n., 79, 86
  bonds 45, 47, 60
  conditional 44, 70 n.
  unconditional 87
  VAR 61–5
  wealth 83
variance–covariance 34, 101, 116
  conditional structures 93–5, 96–9,
    124
  matrix 83, 85, 86, 115
variation 49
  predictable 43, 44, 45
Viotti, S. 3, 7 n.

Wall Street Journal 61
Wallace, N. 25
Walrasian price adjustment rule 73 n.
wealth 32–3, 34 n., 69, 83, 86, 127
  changes, from induced saving 122

Index

dependent on price of capital 120
effects on 79
household 76, 89, 121, 125 n.
initial 73
net, bonds as 18
private, aggregate 11
total 88, 100, 115–18
wealth effects 66, 70, 72
welfare function 26, 27
Werin, L. 7 n., 18 n.
Wooldridge, J. M. 52
work-horse model 67

Yarrow, G. 13
Yeo, E. 111 n.
yields 13, 68, 69, 117, 128
adjustments 10
bond 60, 72, 125, 126; long-term 46, 47, 48, 49–50, 56
capital 119, 121
corporate stock 41, 46, 47, 56, 57

covariance 52, 72
dividend price 39, 123, 135, 138
effects on 36, 48, 49, 51
equilibrium 28, 35, 57, 68 n.
equity 41, 57, 72, 74, 134;
corporate 50, 78
expected 43, 50, 67, 79, 140; cash flows more sensitive to 127; long-term debt 75
market-clearing structure 113, 115, 116
mortgage 125
nominal 62, 76
private 134
real 33, 70 n.
relative 1, 2, 8, 19, 32, 78
structure 27, 114, 124, 137
Treasury bill 131
Yun, K.-Y. 27

Index compiled by Frank Pert